A Practical Textbook of Genetic Engineering in Bacteria

A Practical Textbook of Genetic Engineering in Bacteria

Dr.P.V.G.K.Sarma

Head, Department of Biotechnology

Sri Venkateswara Institute of Medical Sciences and University

Tirupati 517507 Andhra Pradesh, India

Chennai   Trichy   Tirunelveli   NewDelhi

ISBN 978-93-88694-41-4 **MJP Publishers**

All rights reserved   No. 44, Nallathambi Street,
Printed and bound in India   Triplicane, Chennai 600 005

MJP 763   © Publishers, 2020

Publisher: C. Janarthanan

Project Editor: C. Ambica

PREFACE

Genetic manipulation of bacteria is exploited largely in the laboratory because of their simple genetics and as they are easy to grow. These bacteria are predominantly important in producing large amounts of pure recombinant human proteins which can be used in medicine. The first example of this kind was shown in 1978 by Herbert Boyer working in a laboratory in University of California. He expressed human insulin gene in *Escherichia coli.* This product was approved by U.S. Food and Drug Administration for treating diabetes. The ease with which the bacteria could be manipulated has helped researchers in obtaining important molecules which otherwise are very difficult to purify from the organism in large amounts to be used in therapeutic purposes.

The genetic sequences from a broad range of organisms can be easily ligated to a plasmid and transformed into bacteria for both storage and modification. Bacteria are simple to grow, divide quickly, easy to transform, and can be stored at -86 °C more or less indefinitely. The moment a gene is isolated it is very easy to store in bacteria and this provides limitless supply of active material for research. This has led to the development of huge number of recombinant plasmids for manipulating DNA/genes of different prokaryotic and eukaryotic organisms.

The speed with which genetic tools were developed by scientists for understanding gene function, its evolution, variation in the gene sequence is the causative reason for the disease or its role in the pathogenesis or its role in the development of therapeutic products. The majority of DNA exploitation takes place within bacterial plasmids/phages before transfer to another host. In this quest *Escherichia coli* became the organism which helped scientists to exploit, easily manipulate and combine genes to create novel or disrupted proteins and follow the effect on various molecular systems. Researchers have joined genes from bacteria and archaea, paving way to understand the importance of the gene(s) of that organism in different conditions. In this context, the experiments that help to understand the DNA/gene sequence(s) are explained in this book. The experiments mentioned in this book are easy to perform and one can get answers quickly and can purify recombinant proteins rapidly so that they can be used readily for obtaining solutions to various problems. I hope that by following these procedures students can enjoy doing experiments in genetic engineering and learn the beauty of biology.

Dr. P.V.G. K. SARMA

CONTENTS

Chapter 1

CLONING OF DNA FRAGMENTS INTO PLASMID VECTORS

BACKGROUND

Vectors are described as double-stranded DNA molecules having origin of replication of a particular bacterial organism, selection marker for identifying the vector in the host upon transformation and multiple cloning sites comprising of several unique type II restriction endonuclease sites, into which on ligating a DNA insert having same ends can result in the inactivation of the marker helping us to identify recombinants and non-recombinants.

In bacteria these properties are served by: plasmid molecules - an extrachromosomal element, bacteriophages and transposons.

PLASMIDS

All bacteria possess extrachromosmal elements which are covalently closed circular DNA molecules ranging from, ≥1Kb to 500Kb in size, and genes that makes the bacteria to survive in adverse conditions which otherwise would have killed the bacteria. These characters are encoded on the plasmids and they can be: enzymes, toxins, virulence factors and proteins; these structural genes are expressed in the bacteria in those particular environmental conditions. These plasmid molecules also move from one bacterium to another bacterium through horizontal transfer of genetic material from host to the recipient and thus spreading these characters. For example, resistance to β-lactam antibiotics is present on plasmid, similarly vancomycin resistance genes VanA and VanB are encoded on plasmids. This horizontal transfer is mediated by conjugation, transformation and transduction.

Conjugation

Plasmid possess an unique genetic transfer mechanism referred as conjugation, in which one copy of the plasmid DNA from donor is transferred to the recipient through conjugation tube as in Gram-negative bacteria or through surface contacts between bacteria in case of Gram-positive bacteria. Plasmids which transfer genetic material through conjugation phenomenon are referred as **conjugative plasmids** while plasmids that do not possess this feature are referred as **non-conjugative plasmids**. Conjugative plasmids contain two exclusive regions on the plasmid which helps the bacteria to adapt the conjugative phenomenon, these are: i. transfer genes comprising of a cluster of 12 genes and ii. nic/bom regions which facilitate the transfer of DNA through conjugation tube in Gram-negative bacteria while through surface contacts in Gram-positive bacteria.

Mechanism of Conjugation   This is a process in which the expressed proteins from transfer genes recognizes the donor bacteria and synthesizes conjugation tube and allows the plasmid to divide into two copies and one copy passes through the conjugation tube from donor to the recipient and this process is facilitated by the proteins formed from nic/bom regions. This phenomenon is called as **conjugation**. This horizontal gene transfer is a critical phenomenon in which drug resistant plasmids are transferred among bacteria thus, spreading drug resistance across the globe. This phenomenon is usually observed in Gram negative bacteria while in Gram positive bacteria there is no conjugation tube formation, however the transfer is primarily through surface contact between recipient and donor. Here transfer genes traA, traL, traE, traJ, traI, traO and traM facilitate the cleavage of one strand of donor plasmid molecule which forms complex with traJ and traL creates way for the plasmid DNA to move from the pores created by traG (muramidase) and thus pushes the plasmid DNA to move with the help of traO and traM to recipient bacteria.

Transduction

Transduction is the genetic transfer of bacterial DNA from one bacterium (the donor) to another bacterium (the recipient) by a bacteriophage. Transducing bacteriophages are formed in donor bacterial cells during phage development. Transduction is of two kinds: **a) Generalized Transduction**: In this phenomenon the transducing phage carries a random fragment of host chromosomal DNA approximately the same length as that of phage DNA. **b) Specialized Transduction**: In specialized transduction, phages such as Lambda bacteriophage are able to transfer only certain bacterial genes which are near the prophage (when a bacteriophage genome is inserted and integrated into the circular bacterial DNA chromosome and stays as a part of bacterial chromosome; and is also referred as lysogen) insertion site in the bacterial genome. These may be included as a single hybrid DNA molecule of phage genome. These hybrid molecules which are formed by rare aberrant excision of an integrated prophage from a lysogen can transfer the DNA once the phage genome gets integrated into the host bacterial genome. Thus, specialized transduction can be carried only by temperate phages whose lysogens involve reversible integration of the phage genome with that of the host bacterium. This is routinely used to construct mutant bacteria expressing unique enzyme(s), for example using T7 RNA polymerase.

TRANSFORMATION OF PLASMID DNA INTO E.COLI DH5α CELLS

Transformation is a natural phenomenon when the genetic material (DNA) passes through the prevailing medium from donor into recipient organism, and here DNA transfer is entirely dependent on the recipient organism. The acquired DNA expresses heritably inside the recipient organism, this is then referred as **transformation**.

Calcium Chloride Mediated Transformation

Principle

Mendel and Higa originally developed the procedure, which involves exposure of the growing bacteria to a hypotonic solution of calcium chloride at 0 ^{0}C, causing the bacteria to swell (spheroplast formation). DNA added to the transformation mixture forms a DNase resistant complex of hydroxyl calcium phosphate that adheres to the cell surface. The cell during brief exposure to 42 ^{0}C heat pulse can take this complex. After growing for few hours in LB medium in order to revive its cell wall, etc., plating on selective medium screens the transformants. In order to get proper transformation the *E. coli* is grown in a medium enriched with Mg^{2+}, and glucose, fewer amounts of KCl and NaCl. Doubling the tryptone concentration enhances the amino acid metabolism. Transformants are screened based on the controls used: the negative control in which the revived competent cells are plated without any selection pressure. This ensures two things one about the contamination of other microorganisms and the other revival of cell wall of the bacteria. The second control is known as positive control in which revived competent cells are plated with selection pressure. This gives us the information that the revived competent cells lack the genes whose expression will help in the removal of selection pressure. The transformants are screened on the plates containing the selection pressure which ensures entry of plasmid DNA in to the host bacteria.

In case of recombinant plasmids (carrying a foreign DNA) the transformants are screened based on the following controls

Table 1   Different controls used to screen recombinant clones

Control	Selection pressure	Growth
Insert control	Ampicillin+ve, IPTG and X-gal	-ve
Vector control	Ampicillin +ve, IPTG and X-gal	+ve
Ligation control	Ampicillin +ve, IPTG and X-gal	+ve

Control	Selection pressure	Growth
Positive cell control Only competent *E. coli* DH5α bacteria were plated	Ampicillin +ve, IPTG and X-gal	-ve
Negative cell control Only competent *E. coli* DH5α bacteria were plated	No selection pressure	+ve

Based upon these controls whether the cloning is successful or not can be ascertained in the test plate. The number of colonies in the test plate must be at least five to six folds higher than present in the ligation control to accentuate successful formation of recombinants. We calculate the transformation efficiency as follows:

$$\text{Transformation efficiency} = \frac{\text{No.of transformants (colonies) x Final volume at recovery (ml)}}{\mu\text{g of plasmid DNA x volume plated (ml)}} \quad \text{eq.1}$$

Materials

a) 1M $CaCl_2.2\,H_2O$ or TB

Transformation buffer (TB): MES [2-(N-morpholino) ethane-sulfonic acid] buffer contains: 10mM MES pH 6.3, 45mM Mn-$Cl_2.4H_2O$, 10mM $CaCl_2.2H_2O$, 100mM KCl, 3mM Hexaminecobalt chloride, prepare in autoclaved distilled water. After all the salts are dissolved sterilize by Millipore filtration (filters anything between 0.45μM to 0.2μM pore size).

b) Autoclave distilled water

c) SOB (Super optimal broth) - Media

d) LB-Agar.

e) 2M Glucose (sterilize by Millipore filtration)

f) 2M $MgCl_2.6H_2O$ (2M $MgCl_2.6H_2O$: This solution is made by dissolving 19 g of $MgCl_2$ in 90 ml of deionized H_2O. Adjust the volume of the solution to 100ml with deionized H_2O and sterilize by Millipore filtration).

g)   100mg/ml Ampicillin dissolved in autoclaved distilled water

h)   1M IPTG solubilized in autoclaved distilled water.

i)   50mg/ml X-Gal dissolved in N N' dimethyl formamide.

SOB Medium

To 950 ml of deionized H_2O, add:

Bacto tryptone	20 g
Bacto yeast extract	5 g
NaCl	0.5 g

Shake until the solutes have dissolved. Add 10 ml of a 250mM solution of KCl. (This solution is made by dissolving 1.86 g of KCl in 100 ml of deionized water). Adjust the pH to 7.0 with 5N NaOH (0.2 ml). Adjust the volume of the solution to 1 litre with deionized H_2O. Sterilize by autoclaving for 20 minutes at 15 lb / sq in on liquid cycle.

SOC (Super Optimal broth with Catabolite repression) Medium: SOC Medium is identical to SOB medium, except that it contains 20mM glucose and 20mM of $MgCl_2$. Inoculate the media with 0.5ml of mid-log phase bacterial culture and allow it to grow until OD540nm is between 0.5 – 0.6.

Luria - Bertani (LB) Medium

Bacto tryptone	10g
Bacto yeast extract	5g
NaCl	5-10g
dH_2O	1 litre

Adjust the pH to 7.0 with 10 N NaOH (0.4ml/litre).

LB agar plates: 15 g of Bacto agar/litre LB media.

10% Glycerol: 10ml of autoclaved glycerol dissolved in 70ml of autoclaved distilled water and after dissolution the volume is made up to 100ml with autoclaved distilled water.

Method I. (CaCl$_2$. 2H$_2$O mediated transformation)

1. Streak *E. coli* DH5α cells on LB-agar plate and incubate at 37 ^{0}C for overnight. Pick a single colony from the plate and grow up to OD$_{540}$ = 0.6 (10 hrs growth).

2. Inoculate 50 ml of this culture to 100 ml of SOB media containing 20mM of glucose and 20mM of MgCl$_2$.6H$_2$O and grow at 37 ^{0}C for 5 hrs with constant shaking at 250 rpm (such that O.D$_{540}$ = 0.5).

3. Leave the grown *E. coli* DH5α culture on ice for 30 minutes. Pellet the cells by centrifuging at 7,000 rpm for 10 minutes at 4^{0}C.

4. Dilute the stock 1M calcium chloride to 0.1M CaCl$_2$. 2H$_2$O and chill it by keeping it in ice and suspend the pellet in 10 ml of chilled 0.1M CaCl$_2$. 2H$_2$O, or in TB. Leave it in ice for 30 minutes. *[Note: This is 1/10 volume of the bacterial culture volume]*.

5. Centrifuge the cells at 7,000 rpm for 10 minutes at 4^{0}C. Discard the supernatant very carefully as the pellet would be loose and suspend the pellet in 1 ml of ice cold 0.1M CaCl$_2$. 2H$_2$O or TB. At this stage these cells are called as competent cells and they can be stored in 25% v/v glycerol (750µl competent cells and 250µl 100% sterile glycerol) and kept at -20 ^{0}C.

6. Take 200 ml of the competent cells in a sterile 1.5ml microcentri-fuge tube and add 10 ml of the plasmid (containing 1 to 10ng of the DNA) and leave it on ice for 30 minutes. (In case of cloning of a DNA fragment or PCR product or cDNA or genomic library preparation prepare 5 tubes containing 200 ml of the competent cells and to that add one DNA control mentioned in Table 1 such that you have four controls and one tube contains the ligation mixture).

7. Keep the above tubes at 42 ^{0}C for 90 seconds and then immediately leave them on ice for 2 to 3 minutes.

8. Add 800 ml of LB media and grow at 37 ^{0}C for 1hr with constant shaking at 250 rpm.

9. Plate LB-Agar plates with 100mg/ml ampicillin, and incubate at 37 ^{0}C for overnight, following day screen the plates to establish the successful transformation taking into consideration of controls (Table 1 and eq.1). Once the recombinants are ascertained, prepare for

screening the desired clone. Grow the positive clones in LB broth containing 50mg/ml ampicillin and isolate the plasmid by alkaline lysis method.

Alkaline Lysis Method

Principle

The principle behind isolation of plasmid DNA from *E. coli* is first hydro-lyzing the cell wall with lysozyme which breaks the N-acteyl muramic acid and N-acetyl glucosamine bond present in the peptidoglycan layer (the main component of cell wall). The bacteria is then incubated in very high alkaline solution containing anionic detergent (Sodium dodecyl sulfate) for brief period of time which causes disruption of cell membrane, hydrolysis of RNA and denaturation of DNA. The solution in the tube is neutralized using potassium acetate at pH 4.8; this causes immediate precipitation of detergent along with hydrolysed chromosomal DNA, hydolysed RNA, proteins and other components of bacteria, while plasmid DNA escapes due to its supercoiled nature and remains in the aqueous phase which can be collected by centrifugation.

Materials

a. TEG Buffer

    20% Glucose

    50mM Tris-HCl, pH 8.0

    50mM EDTA.

b. Lysis Buffer

    1% SDS

    0.2N NaOH

    Note: For preparing 10 ml of lysis buffer. Take 8.8 ml of sterile distilled water and to that add 0.2 ml of 10N NaOH and 1 ml of 10% SDS. (Note: Do not add SDS first as it will precipitate when NaOH is added to the solution).

c.   3M Potassium acetate pH 4.8: Dissolve 29.44 g potassium acetate in 25 ml of distilled water and adjust the pH to 4.8 with glacial acetic acid. Finally make up the volume to 100 ml of distilled water and sterilize by autoclaving at 15 lb / sq.in for 15 minutes.

d.   RNase. Dissolve 10 mg / ml pancreatic RNase in 50mM Tris-HCl pH 7.5 containing 0.15M NaCl. Keep the above solution in boiling water bath for 10 minutes, this is to destroy the DNase function.

e.   Isopropanol (ice cold)

Chill the isopropanol by keeping it in ice before using.

f.   70% Ethanol

Mix v/v 70 ml of 100% ethanol and 30 ml of sterile distilled water.

g.   10 mg / ml Lysozyme solution. Dissolve 10 mg of lysozyme solution in 1 ml of TEG buffer and leave it on ice.

h.   TE Buffer

10mM Tris-HCl

1mM EDTA, pH 8.0.

Method

1.   Grow E .coli5α containing plasmids pUC 18 or pUC 19 or pQE-30 in 50µg / ml ampicillin at 37° C with a constant shaking at 150-200 rpm up to late log phase (OD540= 0.8 or 0.9)

2.   Take 1.5 ml of the above grown culture in a sterile microcentrifuge tube and centrifuge the cells at 7,000 rpm for 90 seconds at room temperature in a microcentrifuge.

3.   Discard the supernatant and add 0.5 ml of TEG buffer and suspend the culture and centrifuge the cells at 7,000 rpm for 90 seconds at room temperature in a microcentrifuge (in case of Gram positive bacteria repeat this step thrice).

4.   Discard the supernatant and suspend the pellet in 150µl of ice cold lysozyme solution and leave it in room temperature for 10 minutes (in case of Gram positive bacteria incubate at 37° C for 120 minutes).

5. To the same above mixture add 50 µl of 10mg/ml RNase and leave at 37° C for 30 minutes.

6. Now add 300 µl of lysis buffer and mix it carefully such that clear transparent solution is formed and leave it on ice for 10 minutes.

7. Now add 250 µl of 3M potassium acetate buffer pH 4.8 slowly and leave it on ice for 30 minutes.

8. Centrifuge at 10,000 rpm for 10 minutes at 4° C, discard the pellet and to the supernatant add 1:1 (v/v) of ice cold isopropanol mix by gentle vortexing and leave it on ice for 2-3 minutes and centrifuge at 10,000 rpm for 10 minutes at 4° C.

9. Discard the supernatant and add 0.5 ml of 70% ethanol and centrifuge at 10,000 rpm for 10 minutes at 4° C. Discard the supernatant and air dry the pellet.

10. Dissolve the pellet in 50µl of TE buffer. Analyze the plasmid DNA by running in one percent agarose gel.

Method II: Electroporation Mediated Transformation

Principle

All the cell membranes including bacterial membranes are bilayer structures and across this membrane all ions and molecules are transported as per concentration gradients. The aqueous pore formation starts with the infiltration of water molecules into this bilayer membrane which causes reorganization of the adjoining phospholipids facing their polar heads towards the water molecules thus creating temporary unstable pores even in the absence of any external electric field. On application of external electric field, it causes polarization of the membrane which results in the formation of an induced transmembrane voltage. With the infiltration of water molecules increases the rate of pore formation, resulting in a large number of pores generated in the bilayer per unit area and per unit of time, which are more stable than formed in the absence of external electric field. The pure DNA devoid of any salts can easily be trapped inside these pores and easily can be pushed into the cytoplasm where it perpetuates and stabilizes. Basically, here the size of the DNA is not the criteria for the transformation as observed in calcium mediated transformation but it is the purity of the DNA that is the main criterion.

The principle of operation of the electroporation instrument is the creation, within the confines of a series of conductive needle-electrodes arranged in a six-needle hexagonal array, of a pulsed electric field of sufficient strength to induce a transient increase in the permeability of membranes of the cells enclosed within the field. The electroporation instrument supplies a series of precisely timed, short duration pulses of a very specific voltage to pairs of the applicator needle-electrodes and automatically sequences these pulses among the pairs of opposed needles comprising the needle array. This "sequencing" of pulses assures complete and uniform coverage of the electric field within the needle array and consequently complete and uniform electroporation of the cell membranes.

Method II. (Transformation by Electroporation)

1. Each *E. coli*DH5α strain was grown at 37 ^{0}C in 50 ml of LB broth until the culture reached mid-log-phase growth as above.

2. Cultures were chilled on ice for 10 min and centrifuged at 7,000 rpm for 15 minutes at 4 ^{0}C.

3. The supernatant was removed and the cells were resuspended in 50 ml of ice-cold distilled water.

4. The bacterial suspensions were centrifuged again at 7,000 rpm for 15 minutes at 4 ^{0}C. The supernatant was discarded and the cells were resuspended in 50 ml of icecold 10% glycerol.

5. The centrifugation step was repeated once again and the cells were resuspended in 2 ml of ice-cold 10% glycerol.

6. Cells were aliquoted in a sterile microcentrifuge tubes (50 µl per tube) and stored frozen at -80 ^{0}C until use.

7. For transformation take one aliquot of the stored tube and thaw it by keeping in ice for 30 minutes and add clean DNA preparation (devoid of all salts).

8. Cells and DNA were electroporated at 1250 V by using 1mm-gap electroporation cuvette and Eppendorf electroporator as usual.

9. Add 800 ml of LB media and grow at 37 ^{0}C for 1hr with constant shaking at 250 rpm.

10. Plate LB-Agar plates with 100mg / ml ampicillin, and incubate at 37 ^{0}C for overnight, following day screen the plates to ascertain successful transformation considering the controls (Table 1) and eq 1. Once the recombinants are determined prepare for establishing the desired clone.

PARTIAL GENOMIC LIBRARY CONSTRUCTION

Principle

Construction of a genomic library involves: a) isolation of total genomic DNA from the organism, b) digestion of the genomic DNA with type II restriction endonucleases (type II RE) - either complete digestion or partial digestion. For complete digestion the genomic DNA is incubated with type II RE for one hour at 37 ^{0}C and this step is repeated for at least two to three times, this is because the units of the type II RE are defined using a particular DNA where restriction sites for that type II RE are fixed and in case of genomic DNA it is difficult to know the number of sites of that type II RE in that genomic DNA; therefore, it is always advisable to repeat the steps at least two to three times. Also, type II RE are mainly thermolabile enzymes, if large amount of enzyme is added in the beginning itself many type II RE molecules may get inactivated and the desired results may not be achieved. The generated fragments are electrophoresed and desired size of fragments is either eluted from the agarose gel using standard phenol/chloroform method or they may be electroeluted. To achieve this the digested genomic DNA fragments are electrophoresed in 1% agarose gel under submerged conditions, and to know the correct size of the DNA in one well, known molecular size markers are also electrophoresed. The size of DNA fragments to be eluted are marked and agarose containing the fragments is cut and placed in a dialysis bag whose pore size is less than the smallest fragment to be eluted from the agarose gel piece. One side of the dialysis bag is sealed with clamp and to this dialysis bag add 10mM Tris-HCl pH 8 and 10mM EDTA buffer is added two to three times of the gel volume and seal the other end with clamp such that no buffer comes out of the bag, now perform electrophoresis keeping the dialysis bag in the cathode. During the electrophoresis monitor the dialysis bag with hand UV-torch by observing the disappearance of ethidium bromide in the agarose gel which means DNA fragments moved into the buffer from the gel. Carefully take out the buffer containing DNA fragments concentrate by adding ice cold isopropanol. Centrifuge the mixture and collect the DNA as precipitate and suspend it in minimum volume of 10mM Tris-HCl pH 8 and 1mM EDTA buffer. In case of partial digestion, the genomic DNA is incubated with type II RE and the reaction is stopped at 5 minutes, 10 minutes, 20 minutes, 40 minutes, and 60 minutes time interval by adding EDTA which chelates Mg^{2+} ions and ends the reaction. The digested DNA is electrophoresed in submerged conditions along with molecular size markers and pool of different size fragments are electroeluted from the agarose gel as mentioned above (Figure 1).

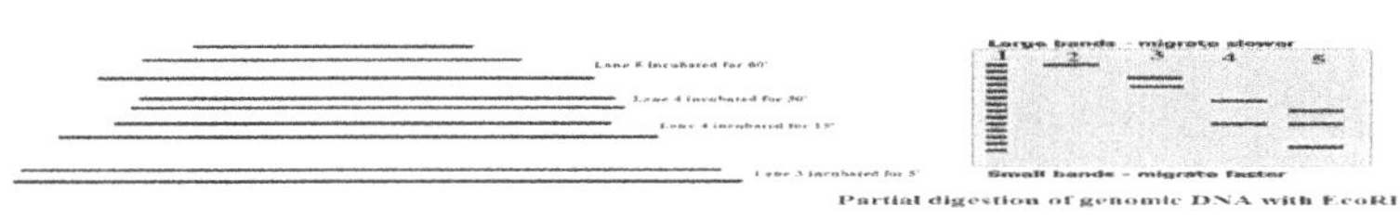

Figure 1 Schematic representation of partial digestion of genomic DNA using type II restriction endonuclease in time dependent incubation and subsequent resolution in agarose gel electrophoresis.

These fragments can be ligated to a vector which is digested with same type II RE or its isoschizomer. The ligation mixture is now transformed into *E. coli* and all the clones thus generated contain different sizes of the genomic DNA digested with type II RE; all these pools of bacteria are called as **genomic library.**

The generated genomic library is cross checked and the desired clone is picked up by colony hybridization which matches with the gene and the desired clone is identified using gene specific probe (Figure 2).

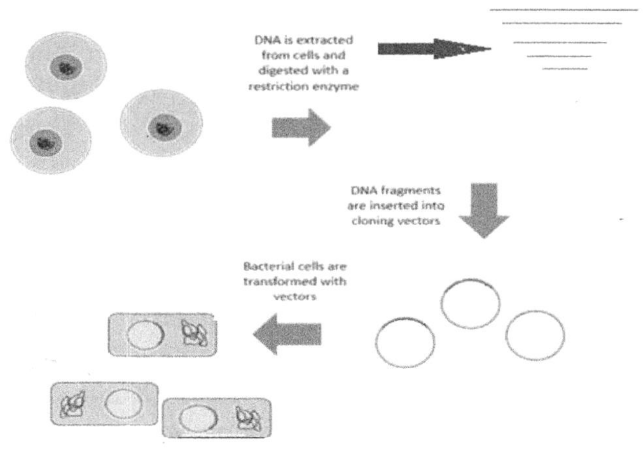

Figure 2   Genomic library construction in pUC 18 plasmids

PARTIAL cDNA LIBRARY CONSTRUCTION

Principle

A cDNA library is a combination of cloned cDNA (complementary DNA) fragments inserted into a collection of host cells, which together constitute some portion of the transcriptome of the organism. cDNA is produced from fully transcribed mRNA found in the nucleus and therefore contains only the expressed genes of an organism. Similarly, tissue-specific cDNA libraries can be produced. In eukaryotic cells the mature mRNA is already spliced, hence the cDNA produced lacks introns and can be readily expressed in a bacterial cell. While information in cDNA libraries is a powerful and useful tool since gene products are easily identified, the libraries lack information about enhancers, introns, and other regulatory elements found in a genomic DNA library.

cDNA is created from a mature mRNA from a eukaryotic cell with the use of reverse transcriptase. In eukaryotes, a poly-(A) tail (consisting of a long sequence of adenine nucleotides) distinguishes mRNA from tRNA and rRNA and can therefore be used as a primer site for reverse transcription.

This has the problem that not all transcripts, such as those for the histone, encode a poly-A tail.

mRNA extraction   Firstly, the mRNA is obtained and purified from the rest of the RNAs. Several methods exist for purifying RNA such as trizol extraction and column purification. Column purification is done by using oligomeric dT nucleotide coated resins where only the mRNA having the poly-A tail will bind. The rest of the RNAs are which are unbound to the column come out as flow through. The bound mRNA is eluted by using eluting buffer which breaks the hydrogen bond formed between 3' poly-A tail of mRNA and oligodT bound to column. Thus eluted total mRNA is concentrated by isopropanol and used in the synthesis of complementary DNA.

cDNA synthesis   The ability of reverse transcriptase to synthesize DNA from an RNA template, allow researchers to study the expression levels of a gene in different conditions in the cell (using real time PCR) and also similar molecular approaches used for DNA manipulations can be employed to understand the subtle functionalities of the genes. cDNA generated by reverse transcription can be amplified using polymerase chain reaction (PCR). The combination of reverse transcription and PCR (RT-PCR) allows the detection of low abundance RNAs in a sample. In the first step of the PCR process, the cDNA is denatured by heating to 95°C, which disrupts the hydrogen bonds between complementary strands, yielding single-stranded molecules. The temperature is then lowered in order to allow primers complementary to the sequence(s) of interest to anneal. The DNA polymerase included in the reaction will then begin DNA synthesis. At this point, the temperature is raised to the optimal activity temperature of the DNA polymerase (usually 72°C) to synthesize a new strand complementary to the template. The process of denaturing, annealing, and extension can be repeated multiple times, with a two-fold increase in the amount of DNA molecules with each cycle. Because PCR can selectively amplify a template, it is an important method for detecting specific nucleic acid molecules in a particular cell or small populations of cells (Figure 3). PCR products can be used in many downstream applications, such as cloning in to plasmid vectors, which further can be used directly for sequencing using dideoxy method of DNA sequencing or next generation sequencing.

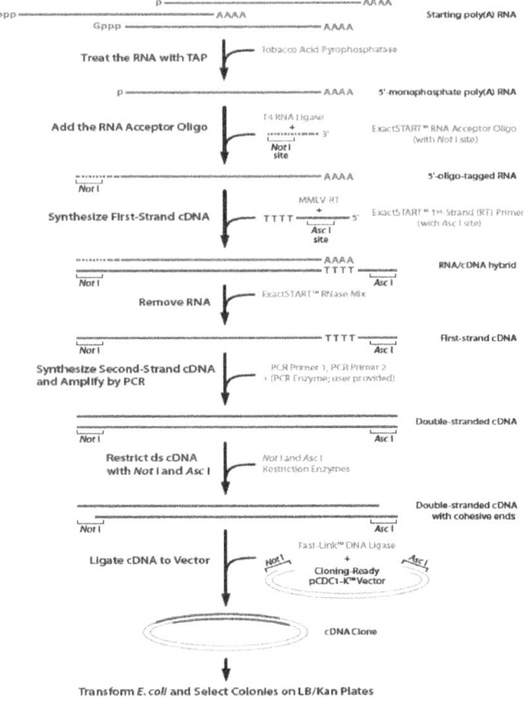

Figure 3 cDNA library construction in plasmid vectors

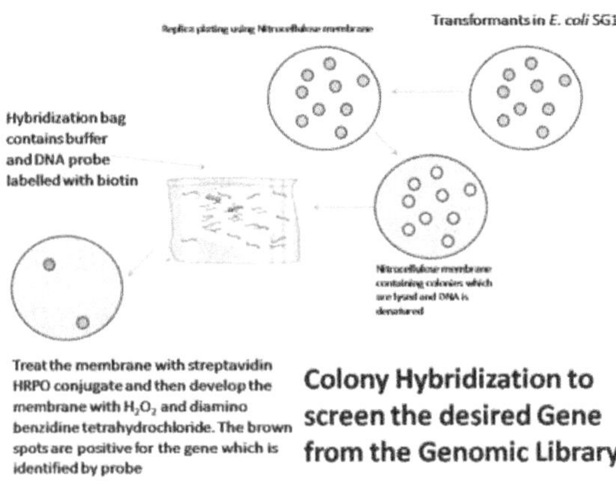

Figure 4 Construction and screening of genomic/cDNA library in plasmid vectors using colony hybridization technique

Screening a genomic library/cDNA library prepared in a plasmid using nucleic acid probe is called as **colony hybridization**. After successful transformation resulting in the formation of recombinants the test plates are screened for desired clone using appropriate probe (Probe is prepared having complete or partial sequence homology with the desired gene). First, plates are kept for 30 min at 4^0C. The plates are then overlaid with same size of nitrocellulose membrane (NCM). Using a sterile needle pierce the agar plate through NCM on the top, bottom, right and left of the plate and label at back of the plate using a marker. Peel the NCM from the plate carefully and incubate the plate at 37^0C for overnight, while placing the NCM on pool of NaOH such that colonies on NCM faces upside and clear surface of NCM are kept on the pool. This is to lyse the bacteria and denature DNA (plasmid as well as chromosomal DNA) and hybridize the colonies with radiolabeled or non-radioactive probe. Similarly, the NCM is treated with 0.05MTris-HCl buffer of pH 7.5. The NCM is either exposed to X-ray film or developed adding suitable substrate to the NCM. On the X-ray film or on the NCM keep the plates exactly matching the wholes that were created on the NCM. Pick up the positive clone and grow it in a liquid broth with the selection pressure added in it. Isolate the plasmid and confirm the presence of insert either by PCR or by restriction analysis (Figure 4).

Materials

A. Restriction endonucleases digestion of Chromosomal DNA

1. *Sau3A digestion of chromosomal DNA*

    10X reaction buffer as supplied by the manufacturer

    20 Unit/μl of Sau3A or as defined by the manufacturer

    0.5M EDTA, pH 8

    50X TAE buffer pH 8.3

    Low EEO agarose

    Ice cold isopropanol

    70% ethanol: 30 ml autoclaved distilled water and 70 ml 100% ethanol.

    Chromosomal DNA

Method

1. Take 5µg (25µl) of DNA isolated from the organism in a sterile 1.5ml microcentrifuge tube.

2. Add 50 µl of 10X reaction buffer

3. Add 425 µl of autoclaved distilled water

4. 2µl of Sau3A (20 Unit/µl)

5. Take 5 sterile 1.5ml microcentrifuge tubes and label them as 5', 10' 20' and 40' and add 50µl of 0.5M EDTA.

6. Incubate the above mixture (steps 1 to 4) at 37 ^{0}C, exactly after 5' pipette 100µl of the reaction mixture and place it in 5' labelled 1.5 ml microcentrifuge tube and vortex briefly and leave at room temperature.

7. Similarly, pipette out 100µl after 10', 20' and 40' and transfer it to the respective tubes containing EDTA.

8. After 60 minutes to the remaining 100µl add 50µl of 0.5M EDTA vortex and leave at room temperature

9. Perform 1% agarose gel electrophoresis along with Supermix DNA molecular size ladder and undigested DNA in 1X TAE buffer under submerged conditions.

10. Record the gel and cut the desired range of DNA fragments gel and electroelute the fragments in 10mM Tris-HCl, 10mM EDTA pH 8.0 buffer present in the dialysis bag. Perform electroelution for 45 minutes to 60 minutes by keeping the bag at cathode.

11. Take out the buffer containing DNA fragments and add equal volume of ice cold isopropanol and mix the solutions briefly by vortexing, following this centrifuge the mixture at 10, 000 rpm for 10 minutes at 4 ^{0}C.

12. Discard the supernatant slowly and add 500µl of 70% ethanol. Vortex briefly and perform the centrifugation as mentioned in step 11.

13. Discard the supernatant very slowly and air-dry the pellet

14. Once all the alcohol is evaporated suspend the pellet in 100µl of TE buffer and use it for the construction of genomic library.

2. *BamHI digestion of pUC 18 vector*

1. 10X reaction buffer as supplied by the manufacturer

2. 20 Unit/µl of Bam HI or 20 Unit/µl of Sma I or as defined by the man-ufacturer

3. 0.5M EDTA, pH 8

4. 50X TAE buffer pH 8.3

5. Low EEO agarose

6. Ice cold isopropanol

7. 70% ethanol: 30 ml autoclaved distilled water and 70 ml 100% ethanol

8. Plasmid pUC 18 DNA 1µg

9. Calf intestine alkaline phophatase (CIAP) (10Units/µl)

10. Solution B: Chloroform:isoamylalcohol mixed in 24:1 (v/v).

Method

1. Take 1µg (10µl) of pUC18 plasmid DNA in a sterile 1.5ml microcentri-fuge tube.

2. Add 5 µl of 10X reaction buffer

3. Add 84 µl of autoclaved distilled water

4. Add 1µl of Bam HI (20 Unit/µl) and incubate at 37 ^{0}C for 1 hour and repeat this step two more times.

5. Add 1µl of CIAP (10 Unit/µl) to the above solution and incubate at 37 ^{0}C for 1 hour and repeat this step two more times.

6. In case of Sma I enzyme digestion: add 1µl of Sma I (20 Unit/µl) and incubate at 25 ^{0}C for 1 hour and repeat this step two more times. (This is for blunt end ligation of cDNA with vector DNA)

7. Add equal volume of solution B and vortex for 2 minutes at room tem-perature and perform centrifugation at 10, 000 rpm for 10 minutes at 4 ^{0}C.

8. Take out the clear aqueous phase in a sterile 1.5ml microcentrifuge tube and add equal volume of ice cold isopropanol and vortex briefly. Perform the centrifugation at 10, 000 rpm for 10 minutes at 4 ^{0}C.

9. Discard the supernatant very slowly and add 250 µl of 70% ethanol and vortex for 2 minutes and perform centrifugation at 10, 000 rpm for 10 minutes at 4 ^{0}C.

10. Discard the supernatant very slowly and air-dry the pellet.

11. Dissolve the pellet in 100µl of TE buffer (10mMTri-HCl and 1mM EDTA pH 8).

3. *Complementary DNA synthesis*

i. *Total mRNA isolation*

Materials

1. DEPC treated distilled water: The distilled water is made RNase free by treating with diethyl pyrocarbonate (DEPC). Add 1 ml of DEPC to 1 litre of distilled water and incubate at 37 °C for 120 minutes; this is to allow DEPC to inactivate ribonucleases. Destroy DEPC by auto-claving at 15 lb / sq.in for 20 minutes and use this water for all buffer preparations, electrophoresis, cleaning of glassware and plastic ware and dissolving RNA.

2. Preparation of glass ware and plastic ware: Soak all glassware in chromic acid for overnight and following day rise in large volumes of tap water and soak for a minute in mild detergent and followed by gentle scrubbing wash it with large volumes of running tap water and follow it up with distilled water. Dry the glassware and soak in DEPC treated water for 1h at room temperature and then bake at 180 ^{0}C for 3 hrs in hot air oven. Treat all plastic ware with chloroform saturated with distilled water and immediately soak in DEPC treated water for 1 hour and autoclave at 121 ^{0}C for 20 minutes.

3. 2X column loading buffer: 40mM Tris-HCl (pH 7.5), 2mM EDTA, 1mM NaCl and 0.2%SDS (w/v)

4. Elution Buffer: 10mM Tris-HCl (pH 7.5), 0.05% SDS and 1mM EDTA

5. Ice cold isopropanol

6. 70% ethanol: 30ml DEPC treated water and 70 ml absolute ethanol

7. 5M NaCl

8. 10N NaOH

Method

1. Suspend 0.5g to 1g of oligo dT cellulose in 0.1N NaOH

2. Pour the column of oligo dT cellulose (0.5ml to 1ml) in a DEPC treated Dispo column or a Pasteur pipette, plugged with glasswool and sterilized by baking for 3h at 180 ^{0}C. Wash the column with column volumes of sterile DEPC treated water.

3. Wash the column with sterile 1X column loading buffer (dilute from 2X stock using sterile DEPC-treated water) until the pH of effluent is< 8, use pH paper for this measurement.

4. Dissolve the total RNA isolated using earlier methods in sterile double distilled water, and heat the solution at 65 ^{0}C for 5 minutes. Cool the solution at room temperature and quickly add the 1 volume of 1X column loading buffer.

5. Apply the solution of RNA to the column and immediately collect the fractions of 1ml. When all the RNA has entered the column, wash the column with 1 column volume of 1X column loading buffer while continuing to collect the flow through till the A260 reaches to 0.001.

6. To improve further the RNA collected is heated at 65 ^{0}C for 5 minutes, and reapply to the column and once again collect the flow through.

7. Wash the column with 5-10 volumes of 1X column loading buffer till the A260 reaches to 0.001.

8. Elute the poly (A+) RNA from the oligo dT column with 2-3 volumes of sterile RNA elution buffer. Collect the fractions and pool them.

9. Precipitate the mRNA adding equal volume of ice cold isopropanol and keep in -20 ^{0}C for > 1h and centrifuge at 10, 000 rpm for 15 minutes at 4 ^{0}C. Wash the pellet twice with 70% ethanol and air dry the pellet till all the traces of ethanol is evaporated.

10. Measure the concentration of RNA and analyze the mRNA by running 1.3%formaldehyde gel electrophoresis.

ii. *cDNA synthesis*

Method

1. Mix RNA sample and primer d(T)23VN in two sterile RNase free micro-centrifuge tubes.

2. Total RNA   6 µl (10 ng–1 µg)

   d(T)23VN (50 µM)   2 µl

   Total Volume   8 µl

3. Denature RNA for 5 minutes at 70°C. Spin briefly and put promptly on ice. This step is optional. However, it improves the cDNA yield for long messenger RNAs and GC-rich RNA regions.

4. Add the following components to one tube.

   5X MMuLV Reaction Mix   4 µl (dilute to 1x depending on the manufacturer's composition)

   MMuLV RT (reverse transcriptase) Enzyme Mix 2 µl (Units choose as per the manufacturer's preparation)

   Nuclease free H_2O variable depending on the total RNA 6µl

5. To the negative control tube, add the following:

   5X MMuLV Reaction Mix   4 µl

   H_2O 16 µl

   Incubate the 20 µl cDNA synthesis reaction at 42 ^{0}C for one hour.

   If Random Primer Mix is used, an incubation step at 25 ^{0}C for 5 min is recommended before the 42 ^{0}C incubation.

6. Inactivate the enzyme at 80 ^{0}C for 5 minutes (some protocol performs at 65 ^{0}C for 10 minutes). Dilute reaction to 50 µl with 30 µl H_2O for PCR.

   The cDNA product should be stored at - 20 ^{0}C after the completion of the reaction.

B. Ligation Reaction

Materials

1. 10X T4DNA ligase buffer with 1mM ATP

2. 100U/µl T4 DNA ligase

3. Autoclaved distilled water.

Method

1. In a sterile 1.5ml microcentrifuge tube add 5µl of Sau 3A digested chromosomal DNA fragments.

2. To the above tube add 10µl of Bam HI digested pUC18 DNA

3. Now add 2µl of 10X buffer

4. Add 2µl of autoclaved distilled water

5. Incubate the reaction mixture at 10 ^{0}C for overnight after adding 1µl of T4 DNA ligase, this we call it as ligation mixture.

6. Similarly, prepare ligation control without adding the Sau 3A digested chromosomal DNA fragments, in place of that add 5µl of autoclaved distilled water and incubate the tube at 10 ^{0}C for overnight.

C. Transformation into E. coli DH5α

Transformation of the ligation mixture and ligation control along with other controls as explained in section 1.1 section is carried and the transformants are screened as per the table:

Table 1   Different controls used to screen recombinant clones

Control	Selection pressure	Growth
Insert control	Ampicillin+ve	-ve
Vector control	Ampicillin +ve	+ve
Ligation control	Ampicillin +ve	+ve
Positive cell control	Ampicillin +ve	-ve
Negative cell control	No selection pressure	+ve

Based upon these controls whether the cloning is successful or not can be ascertained in the test plate. The number of colonies in the test plate must be at least five to six folds higher than present in the ligation control to accentuate successful formation of recombinants. We calculate the transformation efficiency as follows:

$$\text{Transformation efficiency} = \frac{\text{No. of transformants (colonies) x Final volume at recovery (ml)}}{\text{µg of plasmid DNA x volume plated (ml)}} \quad \text{eq.1}$$

Once the recombinants are ascertained, we can use them to screen our desired gene using colony hybridization. The following protocol is followed for the screening of genomic library for the isolation of desired sequence or sequencing whole genome for genomics:

Materials

1. 100mM Tris-HCl, pH 7.5

2. 100mM NaOH

3. 20% Dextran Sulfate

4. 10% SDS

5. 20 X SSPE

    Pre-Hybridization buffer

    6 X SSPE

    1% SDS

    5% Dextran sulfate

    100mg /ml heterologous DNA.

    Hybridization Buffer

    6 X SSPE

    1% SDS

    5% Dextran sulfate

    5-10 pmloes of the oligonucleotide biotinylated probe

Method

1. In 30ml volume take 10 pmole of the probe and add 1 U of T4 DNA polynucleotide kinase and incubate at 37 $^{\circ}$C for 15 minutes with 25ml of (1.5 x 10^7 cpm/ml) g[P^{32}ATP] reaction mixture.

2. Use this probe for the hybridization.

3. Take the overnight grown plate and leave it at 4 ^{0}C for 1 to 2 hrs.

4. Now cut the nitrocellulose membrane with the size of the petri plate and place the nitrocellulose membrane on top of the plate such that entire plate is covered with the membrane. Now mark the plate with the help of a sterile needle punch through nitrocellulose membrane and agar plate. Mark at the back of plate with a permanent marker.

5. Carefully lift the membrane such that colonies stick to the membrane and incubate the plate at 37 ^{0}C for overnight.

6. Make a 3ml pool of 50mM NaOH on Saran wrap paper and place the nitrocellulose membrane; such that colonies face upside and the plane side faces the NaOH. Leave it for 3 minutes.

7. Similarly, place the same membrane on a 3 ml pool of 50mM Tris-HCl pH 7.5 for 3 minutes.

8. Bake the membrane in vacuum for 90 minutes at 90 ^{0}C.

9. Place the membrane in prehybridization buffer for 3 hrs at 42 ^{0}C in a hybridization bag and seal the bag in such manner that there are no air bubbles. Incubate the bag at 42 ^{0}C for 3 hrs with constant agitation.

10. Carefully remove the membrane from the bag and add hybridization buffer and seal the bag similarly and incubate at 42 ^{0}C for overnight with constant agitation.

11. Wash the blot as follows:

    2 x 3 minutes in 2 X SSPE / 0.1% SDS

    2 x 3 minutes in 0.2 X SSPE / 0.1% SDS

    These steps are performed in room temperature.

    2 x 15 minutes in 0.16 X SSPE / 0.2 % SDS at 65^{0}C.

<u>NOTE</u>: Here the conditions need to be identified with the GC content of both probe and the formed library. Expose the membrane to X-ray film (KODAK) in the proper cassette and keep it at -70 ^{0}C or at -86 ^{0}C overnight. The following day develop the film in the developing solution (As per the KODAK manufactur-

er's protocol) or in the automated X-ray film developing instrument.

If the probe is made by adding biotinylated probe then perform hybridization in the following manner.

12. Block the membrane in 0.5% Tween-20 in 0.1M Tris-HCl, pH 7.5 / 0.15 M NaCl.

13. Dilute streptavidin-peroxidase conjugate in 0.1M Tris-HCl, pH 7.5 / 0.15 M NaCl and incubate the nitrocellulose membrane in it at 37 ^{0}C for 1hr.

14. Wash the blot with 0.1M Tris-HCl, pH 7.5 / 0.15 M NaCl for 4 times.

15. Develop the blot by adding 20 ml of developing reagent (6 mg of 3, 3' diaminobenzidine tetrahydrochloride in 10 ml of 0.01 M Tris-HCl pH 8.0 and 30 ml of H_2O_2).

If the probe is tagged with fluorescent agent such as FITC or PE observe the membrane under fluorescent microscope. Imaging and analysis can be performed using a xenon arc-lamp-based gel imaging system with 590 ± 10 nm excitation and 620 ± 10 nm emission filters and a 10 seconds exposure time using Alexa Fluor 594 dye-labelled DNA probe.

16. Place the LB agar plate on top of the developed membrane such that the holes match exactly with the markings on the plate. Now pick up the positive clone which exactly matches with the spot and grow it in 5ml LB broth containing appropriate antibiotics and substrates.

17. Streak the grown culture on a fresh LB agar plate containing appropriate antibiotics.

18. If required repeat the colony hybridization protocol or confirm the insert by PCR or by restriction digestion with appropriate restriction endonucleases.

CLONING OF DNA FRAGMENTS INTO EXPRESSION VECTORS

Immunological Screening

If the library is prepared in an expression vector (plasmid or phage), the encoding gene which you have cloned and expressed it can also be screened using polyclonal antibodies raised against native protein. This is called as immunological screening for identification of correct clone. In this step the cloned gene is expressed if it is cloned in any one of the frames, as explained below (Figure 5):

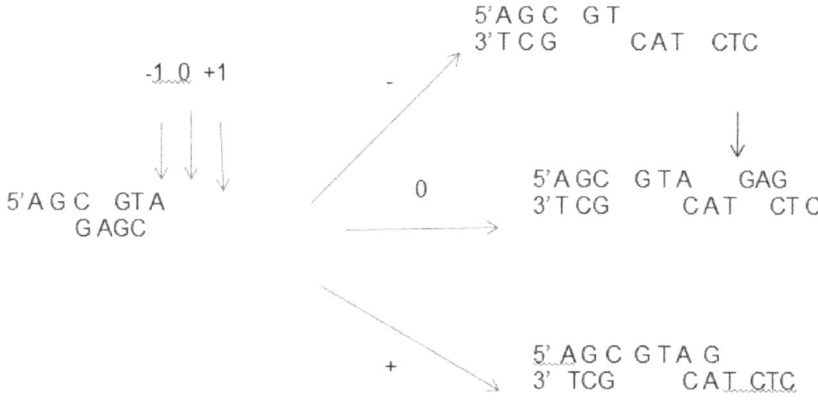

Figure 5   Descriptions "-1", "0" and "+1" indicate cleavages after the second, the third and the first nucleotide of a codon.

Therefore, to achieve cloning in correct frame the vector must be digested with a restriction endonuclease which gives ends that can be ligated with the insert in any one of the frame; for example, if vector is in -1 frame the insert must be in +1 frame, vice versa for +1 cloning while, for 0 frame both vector and insert should be in same frame (Figure). Once the insert is cloned correctly and transformed into suitable host, gene in the insert can be expressed by immersing the NCM with inducer IPTG (isopropyl thio galactoside), in case of lac I and lac operator based expression vector system or with substrate that can trigger the expression of the cloned gene. After keeping at 37 ^{0}C for 6 to 8hrs the NCM is cleanly removed and probed with polyclonal antibodies to check expression in the correct colony or plaque if it is a phage vector. On top of the NCM the plate is placed and correct colony is picked with the help of loop holder and grown in LB broth with appropriate antibiotics and substrates for the propagation of the clone. If it is a plaque use sterile tooth prick and punch the plaque and put it in appropriate *E. coli* culture which is grown up to mid-log phase. Grow further, for at least 5 to 6hrs and then collect the lysate by centrifugation, and identify the expressed protein either by ELISA or by immunological screening (Figure 6).

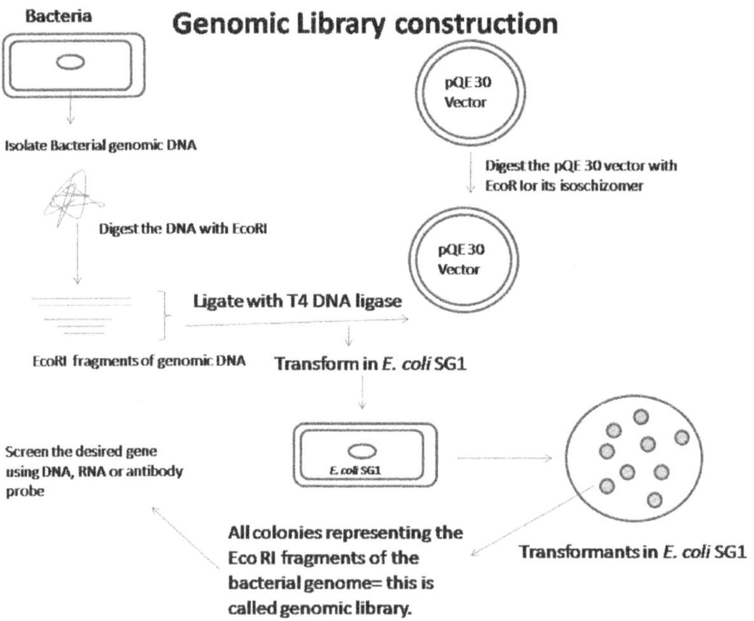

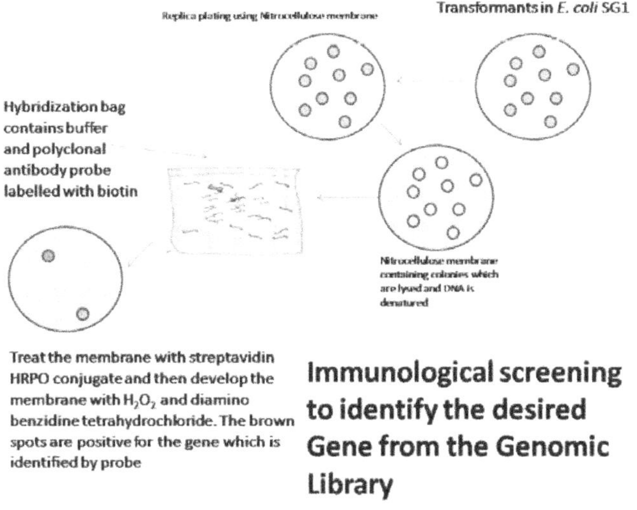

Figure 6 Immunological screening to identify the desired gene from the genomic library. Here we use polyclonal antibodies raised against the native protein which encodes the desired gene.

Immunological Screening of the recombinant plaques

Materials

10 X Tris-HCl

121.1 g Tris base

175.2 g NaCl

Dissolve in 1500 ml distilled water and adjust the pH to 8.0 with concentrated HCl. Then dissolve NaCl in it. Now make up the volume to 2 litres with H_2O.

Wash Buffer

Dilute 1:10 10 X Tris-HCl and then add

0.3 % Tween - 20

0.05% Triton X-100.

Blocking Buffer

Add 5% blotto (5 g non-fat dry milk to 100 ml wash buffer).

Alternatively, use 3% BSA in wash buffer

or 2% gelatin

Method

1. Cut the nitrocellulose membrane with the size of the petri plate and place the nitrocellulose membrane on top of the plate such that entire plate is covered with the membrane. Now mark the plate with the help of a sterile needle punch through nitrocellulose membrane and agar plate. Mark at the back of plate with a permanent marker.

2. Carefully lift the membrane such that colonies are stick to the membrane and incubate the plate at 37^0C for overnight.

3. Incubate the NCM in blocking buffer for 1h at 37 ^{0}C and then perform washing (1 x 5mintues with gentle agitation) 6 times with 30 ml of wash buffer.

4. Incubate the NCM in primary antibody at 37°C for 1 hr. Wash the NCM 6 times with wash buffer.

5. Add secondary antibody (anti rabbit) conjugated to horse radish peroxidase. Incubate at 37 °C for 1 hr followed by 6 times washing with wash buffer.

6. Pour off the last wash buffer and add 20 ml of developing reagent. (6 mg of 3, 3' diaminobenzidine tetrahydrochloride in 10 ml of 0.01 M Tris-HCl pH 8.0 and 30 μl of 30% H_2O_2) and observe the spots specific to the expressed antigen in the clones/recombinants.

7. Place the LB agar plate on top of the developed membrane such that the holes match exactly with the markings on the plate. Now pick up the positive clone which exactly matches with the spot and grow it in 5ml LB broth containing appropriate antibiotics and substrates.

8. If required repeat the hybridization protocol or confirm the insert by PCR or by restriction digestion with appropriate restriction endonucleases.

9. Express the protein from the gene in the positive clones by inducing with appropriate inducer; in case of pQE 30 we induce with IPTG and this induction is always titrated with the concentration of IPTG to get the optimum expression of the protein without killing or effecting the growth of *E. coli.*

10. The expressed protein in the positive clone is released by exposing the grown clone to sonication in a sonicator at 50Hz for 15 to 20 cycles. The bacteria are suspended in sonication buffer (0.1M Tris-HCl pH 8, containing 10% sucrose). The protein is also released from the bacteria using French press. The obtained protein is purified by passing through nickel metal chelate column and pure recombinant protein is screened using immunoblot; if the gene encodes an enzyme perform enzyme assay.

SDS-PAGE

Materials

Lower Gel Buffer

1.5 M Tris-HCl pH 8.8.

0.4% SDS.

Upper Gel Buffer:

0.5 M Tri-HCl pH 6.8

0.4% SDS

30% Acrylamide Stock. Dissolve 29.2 g of acrylamide and 0.8 g N, N'-methylene-bis-acrylamide in about 50ml of distilled water and then make up the volume to100 ml volume. Filter and store the solution at 4 °C in a amber colored bottle.

10% Ammonium persulfate solution: Dissolve 0.1 g of ammonium persulfate in 1 ml distilled water.

SDS-Running Buffer

0.025 M Tris-HCl

0.192 M Glycine pH 8.3

0.1% SDS

2 X Sample Buffer

20% Glycerol

0.125 M Tris-HCl; pH 6.8

0.1% Bromophenol blue

4.6% (w / v) SDS

10% (w / v) 2-mercaptoethanol.

Staining Solution

Distilled water

Acetic acid

Methanol (5:1:4 v/v/v)

0.125% (w/v) Coomassie Brilliant blue R250

Destaining solution

Distilled water

Acetic acid

Methanol (5:1:4 v/v/v)

Method

1. Mix the volume of stock solutions for 15 ml separating gels

S.No		5%	7.5%	10%	12.5%	15%
1	4X Lower Gel Buffer	3.75ml	3.75ml	3.75ml	3.75ml	3.75ml
2	30% Acrylamide stock solution	2.5ml	3.75ml	5ml	6.25ml	7.50ml
3	Distilled water	8.7ml	7.45ml	6.20ml	4.95ml	3.7ml

S.No		5%	7.5%	10%	12.5%	15%
4	*TEMED	0.01ml	0.01ml	0.01ml	0.01ml	0.01ml
5	**APS	0.05ml	0.05ml	0.05ml	0.05ml	0.05ml

*N, N', N' N' Tetramethylethylenediamine

**Ammonium persulfate

2. Mix the volume of stock solutions for 6ml stacker gels

S.No		4%	3.5%
1	Upper Gel Buffer	1.25ml	1.25ml
2	30% Acrylamide stock solution	0.8ml	0.75ml
3	Distilled water	3.89ml	3.94ml
4	*TEMED	0.01ml	0.01ml
5	**APS	0.05ml	0.05ml

*N, N', N' N' Tetramethylethylenediamine

**Ammonium persulfate

3. Wash the glass plates gently with mild detergent and rinse with large volumes of distilled water and air dry.

4. Place the spacers according to the thickness of gel requirement, for example if 0.5cm thickness gel place 0.5cm spacers on one of the gel casting plate and place the other plate and seal the bottom of the plates either with 1.3% agar solution or with vacuum generated (BIORAD apparatus) by placing the plates in casting apparatus.

5. Prepare the reaction mixture taking volumes as explained in step 1 and as per the requirement prepare the percentage of gel. Degasify the solution using a vacuum pump and add APS and TEMED and pour the solution immediately in between the glass plates, make a layer of water saturated isopropanol (isopropanol will help in preventing the dissolution of atmospheric air) and allow it to polymerize.

6. After the formation of separating gel prepare the stacker gel according to the volumes from step 2 and place the clean comb and allow it to polymerize.

7. After the formation of gel take out the comb and prepare the sample by mixing 1:1 (v/v) with 2X sample buffer and keep the sample in boiling water bath for 10-15 minutes. Spin the samples at 10,000 rpm for 30 seconds at 4 °C and load the sample in the gel wells.

8. Perform the electrophoresis at 5V/cm till the bromophenol blue reaches at the end of the gel. If you want to transfer the resolved proteins on the gel into nitrocellulose paper proceed from this step for transfer. After the completion of the transfer, to know if the proteins have completely transferred into NCM, perform the following steps.

9. Take out both the plates along with gel from the electrophoresis apparatus and carefully take out the upper plate with the help of a spatula and scoop the gel in a clean staining chamber and flood with 0.125% CBBR250 stain and leave at room temperature for 3 to 4h. Following staining, destain the gel by dipping in destaining solution till you observe clear protein bands.

10. Calculate the Rf value of the molecular weight markers and sample protein bands. Plot a graph between Rf value of known protein markers and log molecular weight and from the plot the determine molecular weight of the sample protein.

Immunoblot

Materials

Transfer Buffer

-2.2 g CAPS (3-cyclohexyl amino1-propanesulfonic acid)

-Dissolve in 500 ml H_2O adjust the pH to 10.5 and make up the volume to 900 ml. Leave in cold room to pre-chill.

-Just before use add 100 ml of cold methanol.

10 X Tris-HCl

121.1 g Tris base

175.2 g NaCl

Dissolve in 1500 ml distilled water and adjust the pH to 8.0 with concentrated HCl. Then dissolve NaCl in it. Now make up the volume to 2 litres with H_2O.

Wash Buffer

Dilute 1:10 10 X Tris-HCl and then add

0.3 % Tween - 20

0.05% Triton X-100.

Blocking Buffer

Add 5% blotto (5 g non-fat dry milk to 100 ml wash buffer).

Alternatively, use 3% BSA in wash buffer

or 2% gelatin

Nitrocellulose membrane stain

-0.2 g of Amido Black

-90 ml of Methanol

-20 ml of Acetic acid.

Make up to 200 ml with distilled water.

Nitrocellulose membrane destain

- 900 ml Methanol

- 20 ml Acetic acid

- 80 ml dH_2O.

Primary antibody against native protein (Polyclonal) generated in rabbit

Secondary antibody raised against the primary antibody or anti

Methods Resolve the proteins by running SDS-PAGE as described earlier in chapter 2 (2.5 to 2.5.5) of the book Molecular Biology – A Practical Manual and then follow the procedure:

1.   Cut a piece of PVDF (ployvinylidine difluoride, use only for protein sequencing) or nitrocellulose membrane (NCM) to fit the gel. Rinse the membrane with distilled water followed by soaking in the transfer buffer.

2.   Equilibrate the plates of semidry apparatus by soaking two or three absorbent pads in transfer buffer.

3.   In transfer buffer, sandwich gel against NCM and between 2 or 3 sheets of Whatman filter paper. Roll a glass Pasteur pipette gently over the membrane in one direction to make sure that there are no air bubbles between pads and NCM. Place the sandwich in transfer apparatus with NCM towards anode.

4.   Determine the current needed for electrophoretic transfer by using the following formula: $0.8mA/cm^2$ of gel. Transfer is complete in 30 minutes for minigel or 60-90 minutes for full length gel.

5.   The NCM is air dried at room temperature. Cut off the molecular weight marker strip, stain this for 1 hr in NC membrane stain, followed by destaining with changes until background stain eliminated. Stain the gel as usual with coommassiee brilliant blue to observe if any bands are not transferred on to the NCM.

6.   Place rest of NCM in blocking buffer at 37°C for 1 hr with rocking. Wash NCM with (6 times) wash buffer.

7.   Incubate the NCM in primary antibody at 37°C for 1 hr. Wash the NCM 6 times with wash buffer. Add secondary antibody (anti-rabbit) conjugated to horse radish peroxidase. Incubate at 37 °C for 1 hr followed by 6 times washing with wash buffer. Pour off the last wash buffer and add 20 ml of developing reagent. (6 mg of 3, 3' diamino-benzidine tetrahydrochloride in 10 ml of 0.01 M Tris-HCl pH 8.0 and 30 µl of H_2O_2) and observe the bands specific to the antigen. This is called immunoblot technique.

NOTE: When staining and washing membrane always have the protein side up and avoid pouring solutions directly onto the surface of the membrane to prevent breakage of the NCM or PVDF. For large proteins (i.e > 80KD) transfer should be done with < 10% gels for best yields.

Purification of recombinant protein by nickel metal chelate chromatography:

Principle   The genes cloned in vectors like pQE 30 or pET are expressed with their NH_2 terminal fused with 6-H residues. It is well known that Histidine does not disturb the conformation of proteins as they participate poorly in the formation alpha helix structures in the protein, therefore it is very unlikely that proteins properties are disturbed due to the fusion of these 6-H residues. These 6-H residues binds to nickel metal ion through coordinate covalent bonding and is exploited in the purification of recombinant proteins expressed from vectors like pQE 30 or pET. The recombinant proteins present in the cytosolic fractions are passed through column of agarose, or sepharose coupled to nickel ions. The proteins bound to nickel can be easily eluted using buffers containing guanidine hydrochloride or immidazole hydrochloride. The obtained proteins are dialysed overnight to remove guanidine hydrochloride or immidazole hydrochloride and are used directly for protein characterization studies such as enzyme assay or toxin assay or other characteristic features of the protein.

Materials

Nickel metal affinity agarose (Mereck)

Lyse Buffer: 300 mM NaCl, 50 mM NaH_2PO_4, pH to 8.0 with NaOH.

Wash buffer: 50mM NaH_2PO_4, 300mM NaCl, 20mM Imidazole pH 8 using NaOH

Elution buffer: 50mM NaH_2PO_4, 300mM NaCl, 250mM Imidazole

1. To a sterile 1.5ml microcentrifuge tube, 50–100µl of nickel metal affinity agarose gel suspension is added and centrifuged for 60 seconds at 5,000 rpm.

2. Discard the supernatant carefully and 200 µl of Equilibration Buffer without imidazole was added and mixed well.

3. Centrifuge for 60 seconds at 5,000 rpm and discard the supernatant. Add 100 µl of clarified recombinant protein solution (cytosolic fraction) and gently mix for 1 minute using a vortex mixer. Repeat the centrifugation step.

4. The affinity gel is washed two times with at least 500 µl of wash buffer without imidazole. Gently mix the affinity gel for 10 seconds, then

centrifuge for 60 seconds at 5,000. **Save the Wash Buffer solutions for analysis either as a single pool or three fractions.**

5.   The target protein is eluted with 50-100µl of Elution Buffer. After addition of the buffer, mix the affinity gel well by vortexing following this centrifuge for 60 seconds at 5,000rpm.

6.   Repeat steps 4 and 5 to recover more of the protein. *Most of the protein will elute off in the first 50 µl fraction, but some residual protein may be eluted in the second cycle. Save the two fractions as a single pool or separate fractions.*

7.   Analyze all of the fractions by SDS-PAGE to determine if your target protein bound to the affinity gel was eluted. It is useful to perform a Western blot to determine where the histidine containing proteins are fractionated during the purification trial.

CONSTRUCTION OF GENOMIC LIBRARY IN BACTERIAL ARTIFICIAL CHROMOSOME:

Bacterial Artificial Chromosomes

A bacterial artificial chromosome (BAC) is a plasmid DNA constructed using the functional properties of fertility plasmid (or F-plasmid). Usually *E. coli* F-plasmids contain partition genes that help in the smooth distribution of plasmids after bacterial division (which ensures they are maintained at one copy per bacterium). The BACs are capable of carrying insert DNA between 150-350 kbp. Common genes that are present in BAC are:

1.   repE: for plasmid replication and regulation of copy number

2.   parA and parB: for partitioning of F plasmid DNA to daughter bacteria during division and ensures stable maintenance of the BAC (Here, parA and parB containing F plasmids are generated from Hfr strains where F plasmids illegitimately divide in the bacteria giving rise to two molecules one with par A and B and other with par C)

3.   A selectable marker: for antibiotic resistance (Chloramphenicol resistance, *Cm*ʳ); some BACs also have lacZ at the cloning site for blue/white selection

4. T7 and Sp6 promoters: for efficient transcription of inserted genes (Figure 7).

Common gene components

Bacterial artificial chromosome is another cloning vector system in *E.coli* (**pBAC108L**), developed by Melsimon and his colleagues in 1992, have

❑ *HindIII* and *BamHI*: the cloning sites

❑ CmR: the chloramphenicol resistance gene, used as a selection tool.

❑ *oriS*: the origin of replication

❑ *repE*: for plasmid replication and regulation of copy number.

❑ *ParA* and *ParB*: the genes governing partition of plasmids to daughter cells during division and ensures stable maintenance of the BAC.

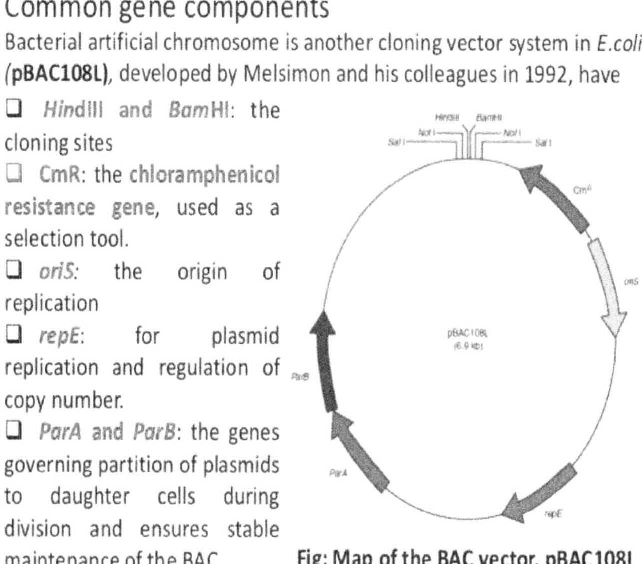

Fig: Map of the BAC vector, pBAC108L

Figure 7   Schematic representation of pBAC plasmid genetic map.

Once an insert is ligated into a BAC, the BAC is introduced into recombination deficient strains of *E. coli* by electroporation. Most BAC vectors contain a gene for antibiotic resistance and also a positive selection marker. The figure to the right depicts a BAC vector being cut with a restriction enzyme, followed by the insertion of foreign DNA that is re-annealed by a ligase. Overall, this is a very stable vector due to the presence of single copy in the host *E. coli* (Figure 8).

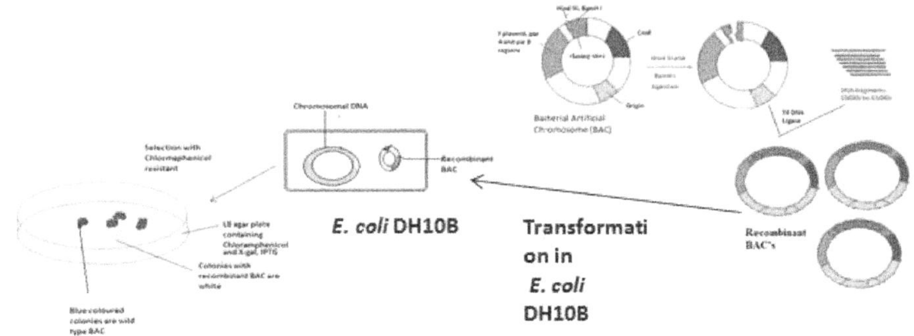

Figure 8   Schematic representations of cloning of insert DNA fragments into BAC vector resulting in generation of partial genomic library in BAC

Materials

1. BAC vector: pBACe3.6 (11.5Kbp)

2. *E. coli* DH10B strain

3. PFGE (Pulsed Field Gel Electrophoresis) apparatus

4. Electroporator (Eppendorf).

5. LB media, LB Agar plates containing 20 μg/ml chloramphenicol

6. SOB media

7. Restriction endonucleases (BamHI or HindIII)

8. *Staphylococcus aureus* ATCC12600 chromosomal DNA or Human chromosomal DNA

9. 5% (w/v) sucrose and chloramphenicol

10. 5% (w/v) sucrose, 100 μg/ml ampicillin, and chloramphenicol

11. BamHI and HindIII restriction endonucleases and 10× buffers (New England

   Biolabs or equivalent)

12. 0.7% (w/v) agarose gels (for standard electrophoresis)

13. Calf intestine alkaline phosphatase (AP; Boehringer Mannheim)

14. 10 mg/ml proteinase K (Boehringer Mannheim) stock solution

15. 95% (v/v) ethanol

16. 1.0% (w/v) agarose solution (ultra-pure, LifeTechnologies; for PFGE system)

17. 6× gel loading buffer, not containing xylene cyanol FF

18. Super mix DNA ladder

19. T4 polynucleotide kinase (New England Biolabs)

20. 30% (w/v) polyethylene glycol (PEG) 8000

21. Dialysis tubing of 3/4-in. diameter, mol. wt. exclusion limit 12,000 to 14,000 daltons (Life Technologies or equivalent)

22. Dialysis clips

23. Alkaline lysis solution: 0.2 N NaOH, 1% (w/v) SDS Prepare before use (take autoclaved distilled water first add NaOH followed by SDS)

24. Precipitation solution: 3 M potassium acetate, pH 5.5, Autoclave, Store indefinitely at 4°C

25. Proteinase K lysis solution: 10 ml filtered 10% N-lauroylsarcosine (sodium salt; Sigma; final 2%) 40 ml 0.5 M EDTA, pH 8.0, 100 mg proteinase K (Boehringer Mannheim; final 2 mg/ml)

    Prepare fresh and use immediately after adding proteinase K.

26. 10 X BamH I endonuclease/methylase buffer (as supplied by the manufacturer),

27. T4 DNA ligase

28. T4 DNA ligase blunt end ligation buffer and cohesive ligation buffer (as provided by the producer).

Method

Test vector DNA

1. Streak *E. coli* DH10B cells containing BAC vector (pBACe3.6) on a LB agar plate having 50µg/ml chloramphenicol and incubate overnight at 37 °C.

2.   Pick single colony and inoculate into a sterile 15-ml screw-cap polypropylene tube containing 5 ml LB medium with 20µg/ml chloramphenicol. Grow cultures overnight at 37 ^{0}C.

3.   Use 1.5 ml of culture to prepare DNA using alkaline lysis protocol. Store the remaining culture at 4 ^{0}C.

Alkaline lysis procedure

i.   Using a sterile toothpick, inoculate a single isolated BAC clone into 5 ml LB medium containing 20 µg/ml chloramphenicol in a 15-ml screw-cap polypropylene tube. Remove toothpick using sterile forceps. Grow overnight (≤16 hr) at 37 ^{0}C in an orbital shaker at 200 rpm.

ii.   Following day centrifuge 5 min at 7,000 rpm at 4 ^{0}C (The temperature of this centrifugation is not critical).

iii. Discard supernatant and resuspend (vortex) pellet in 0.3ml resuspension solution (100mMTris-HCl pH 8, 50mM EDTA and 20% glucose).

iv.   Add 0.3ml alkaline lysis solution and gently shake tube to mix contents. Let it stand at room temperature for ~5 min. Following this keep in ice for 10 min. Note: The appearance of the suspension should change from very turbid to almost translucent.

v.   While gently shaking, slowly add 0.3 ml precipitation solution. Stand tube on ice 30 min. A thick white precipitate of protein and *E. coli* DNA will form.

vi.   Centrifuge at 10,000 rpm at 4 ^{0}C for 15minutes, dispense carefully the supernatant on to a sterile 1.5ml centrifuge tube and add equal volume of ice cold isopropanol. Mix the solution by gentle vortexing, repeat the centrifugation step.

vii. Discard the supernatant and resuspend the pellet in 500µl of 70% ethanol and repeat above centrifugation step.

viii. Discard slowly the supernatant and air dry the pellet and suspend the pellet in 100µl of TE (10mMTris-HCl pH 8 and 1mMEDTA) buffer.

4.   Digest 1µg DNA from the preparation with BamHI, or with Hind III, using the manufacturers' recommended conditions.

5.   Analyze digested DNA in a 0.7% agarose gel.

6.   Isolate the chromosomal DNA from Human Blood as described:

(i)   Erythrocye lysis buffer:0.15M Ammonium Chloride, 1mM Potassium bicarbonate, 0.1mM EDTA.

(ii)   80 % Alcohol

(iii)   10mg/ml Proteinase K

(iv)   RNase Buffer: Pancreatic ribonuclease A, 10 mg/ml, in 0.1Tris-HCl pH 7.5 and containing 150mM NaCl, preheated to 37 ^{0}C for 30 minutes.

(v)   TEN Buffer: in 0.1Tris-HCl pH 7.5 and containing 150mM NaCl, and 50mM EDTA

(vi)   TE Buffer: 10mM Tris and 1mM EDTA

Procedure

a.   Chromosomal DNA from the tissue samples of patients was extracted by salting in and salting out method. 1g of tumour tissue was taken and homogenized by using TE buffer.

b.   The homogenized tissue sample/blood was induced by adding 40 ml of ELB in a 50ml falcon tube and the tube was kept in ice for 30 minutes. It was centrifuged at 3000 rpm for 10 minutes and supernatant was discarded.

c.   The pellet is vortexed (pellet is WBC) and suspended in 10 ml of ELB and mixed gently such that clear solution is observed in the falcon tube. 500 µl of 20 % SDS is added in order to disrupt the cell membrane and other membranes present in different cell organelles. To the above solution 80µl of proteinase K is added and incubated at 37 ^{0}C overnight.

Salting out

i.   On the following day the tube is removed from the water bath and 1 µl of 5 M NaCl is added. To that equal volume of ice cold isopropanol is added slowly (In this step, DNA is precipitated and the precipitate floated in the solution). Thus, precipitated DNA is then removed by centrifugation at 4 ^{0}C for 10 minutes at 10,000 rpm.

ii. The pellet thus obtained was washed with 500 µl of 80% alcohol and centrifuged at 13000 rpm for 10 minutes. This step was repeated for three times to get pure DNA without salts. The pellet was air dried and the suspended in 1 ml of TE buffer.

7. Perform partial digestion of the isolated DNA using either Sau3A or Mbo I and analyse the generated DNA fragments on 1% agarose gel electrophoresis along with molecular size markers.

    i. Incubate 10µg of isolated DNA

    ii. 10 X reaction buffer

    iii. Sau3A or Mbo I 5U/µl

    iv. Autoclaved distilled water so as to make up the entire volume up to 500µl.

    v. Incubate at 37 ^{0}C

    vi. After 5 minutes of incubation please transfer 100µl from the above reaction mixture to a sterile eppendorf (1.5ml) tube containing 25µl of 0.5M EDTA, mix the solution using vortex mixer. Similarly, take out another 100µl and pipette out in another sterile autoclaved 1.5ml tube after 10 minutes. Likewise pipette out 100µl of the reaction mixture in two more tubes and finally add 25µl of 0.5M EDTA to the initial tube.

    vii. Perform 1% agarose gel electrophoresis in 1 x TAE (Tris-Acetate-EDTA pH 8.3) system along with molecular size marker.

8. Cut the agarose containing desired size of DNA fragments and electro elute the fragments.

    a) Place the agarose containing desired fragments in dialysis bag and add 1ml of TE (10mM Tris-HCl, pH 8 and 10mM EDTA) buffer, seal both the sides with clamps and perform electrophoresis keeping the dialysis bag towards the cathode.

    b) After 1h when the DNA fragments from gel must have moved into the buffer stop (by observing the disappearance of orange fluorescence of ethidium bromide from the agarose gel) the electrophoresis and remove thc buffer and precipitate the DNA fragments by adding equal volume of ice cold isopropanol and mix by gentle vortexing. Centrifuge at 10,000rpm for 10 minutes at 4 ^{0}C. Discard the supernatant and add 250µl of 70% ethanol and suspend the pellet by vortexing and repeat the centrifugation step. Discard the supernatant very carefully and air dry the pellet. Suspend the DNA pellet in 20µl of TE buffer.

C)   If chromosomal DNA is from human tissues then to get better results perform pulsed field gel electrophoresis and elute large fragments for the preparation of BAC libraries.

Pulsed Field Gel Electrophoresis

Principle   In pulsed field gel electrophoresis (PFGE) the electric field is altered intermittently perpendicular to the earlier direction with pulse time ranging from 0.1 to 1000 seconds or more. This method can resolve DNA molecules up to ~2Mb. Separation results as the time required for DNA molecule to change direction in response to the changing electric field depends on its size; shorter molecules can reorient more rapidly and can travel in the gel faster than longer molecules.

Materials

Agarose – embedded DNA samples

Rare cutting restriction endonuclease (SfI, BstEHII)

1 mg/ml BSA, acetylated and nuclease free

100 mM Spermidine trihydrochloride

TE buffer

Agarose

Electrophoresis buffer (0.5xTBE or 1x TAE)

6 x loading dye (Composition and preparation is mentioned in method Table 3 and 4)

10mg/ml ethidium bromide

100 mm-petri plates

Water bath for restriction digestion / incubator

Pulsed – field gel electrophoresis apparatus

Constant – voltage power supply,

Pulse generator.

Method

i. Use DNA samples that are embedded in agarose block then place the block in a petriplate and cut into ~25μl (1x4x6mm) pieces that contain ~3μg of DNA ["Handle the sample carefully, the size and shape of bands on the pulsed – field gel will closely reflect the size and shape of block, any variation in the preparation of the sample the bands will not appear in proper fashion but you will see in a smear fashion"].

ii. In a sterile 1.5ml microcentrifuge tube mix the following and kept in ice:

    a. 10 μl 10x restriction buffer

    b. 10 μl 1mg/ml acetylated BSA

    c. 4 μl 100 mm Spermidine trihydrochloride

    d. 50 μl sterile H_2O.

iii. Add 3μg of DNA to the reaction mixture and leave it on ice for 5 min.

iv. Add 1μl (15-20units) of SfI and incubate for 60 min on the ice to allow the enzyme to diffuse into the agarose matrix.

v. Incubate the tube for 5 hr on a water bath at 50 ^{0}C (optimum temperature of SfI), 2 to 3 drops of mineral oil is added to prevent evaporation.

vi. Stop the reaction by diluting with 1 ml of TE-buffer (Dilution and storage in TE buffer allows salts in the reaction mixture to dialyze away from the agarose sample prior to electrophoresis, preventing band distortion); if a second digestion is to be carried then incubate the tube for 30 min at room temperature to permit the buffer from the first endonuclease reaction to diffuse out of the block.

vii.   Store the samples at 4 °C at this temperature one can store sample for several months. For many mapping applications, double digestions are very useful;

viii. TE buffer is removed carefully by centrifuging at 4 °C for 45 seconds and discard the supernatant.

ix.   Repeat steps ii to vi with the second restriction enzyme under conditions suitable for the enzyme.

9.   Take 5µl BamH I digested BAC (pBACe3.6) (50ng) in a sterile 1.5ml microcentrifuge tube and to that add 2µl of 10X cohesive end ligation buffer containing 1mMATP, 2µl of insert (electroeluted DNA fragments (20ng)), 1µl of 100U/µl of T4DNA ligase and 10µl of sterile distilled water. This tube should be labelled as ligation test. Similarly, prepare ligation control in which insert DNA is not present. Incubate both the tubes at 10 ^{0}C for overnight.

10.   Extract the DNA (test) with chloroform and isoamyl alcohol (24:1 v/v) from both ligation control and ligation test tubes by centrifuging at 10,000 rpm for 10minutes at 4 ^{0}C. Collect the aqueous phase in a sterile 1.5ml microcentrifuge tube and precipitate the DNA by adding equal volume of ice-cold isopropanol and centrifuge at 10,000 rpm for 10 minutes at 4 ^{0}C. Discard the supernatant and add 250 µl of 70% ethanol and repeat the centrifuge step. Discard the supernatant slowly and air-dry the pellet. Suspend the pellet in 10µl of TE buffer.

11.   Prepare EP (electro porating) cells (*E. coli* DH10B) following the protocol mentioned under transformation in this chapter. Add 400µl of LB media and grow the cells at 120rpm at 37 ^{0}C in a rotator orbital shaker for 1hour.

12.   Plate the cells in LB agar plate containing 50µg/ml chloramphenicol and incubate at 37 ^{0}C for 15hours. Similarly, prepare for ligation control and observe the number of colonies in the ligation control which must be always 10 to 50 folds less than that observed in ligation test plate indicating successful formation of genomic library in BAC.

13.   To identify the desired clone, perform colony hybridization by using suitable probe against the target sequence.

14. Once the clone is identified, pick the clone and propagate in the presence of 20µg/ml chloramphenicol and isolate the BAC following the procedure described above.

15. Perform Pulsed field gel electrophoresis (PFGE) as described earlier; perform along with BAC DNA as control.

References

1. Das S, Das HR, (2015) Cloning and Transformation In *Microbial Biotechnology- A laboratory Manual for Bacterial Systems*, page number 35-44. DOI 10.1007/978-81-322-2095-4_2

2. Garcia PA, Ge Z, Kelley LE, Holcomb SJ, Buie CR. High efficiency hydrodynamic bacterial electrotransformation. *Lab Chip*. 2017 Jan 31;17(3):490-500.

3. Tu Q, Yin J, Fu J, Herrmann J, Li Y, Yin Y, Stewart AF, Müller R, Zhang Y. Room temperature electrocompetent bacterial cells improve DNA transformation and recombineering efficiency. *Sci Rep*. 2016;6:24648

4. Neumann, E; Schaefer-Ridder, M; Wang, Y; Hofschneider, PH (1982). "Gene transfer into mouse lyoma cells by electroporation in high electric fields". *The EMBO Journal*. 1 (7): 841–5. PMC 553119free to read. PMID 6329708.

5. Sugar, I.P.; Neumann, E. (1984). "Stochastic model for electric field-induced membrane pores electroporation". *Biophysical Chemistry*. 19 (3): 211–25. doi:10.1016/0301-4622(84)87003-9. PMID 6722274.

6. Potter, Huntington; Heller, Richard (2003-05-01). "Transfection by Electroporation". *Current Protocols in Molecular Biology*. CHAPTER: Unit–9.3. doi:10.1002/0471142727.mb0903s62. ISSN 1934-3639. PMC 2975437free to read. PMID 18265334.

7. Sambrook J, Russell DW, *Molecular cloning. A laboratory manual*, Cold Spring Harbor Laboratory Press, Cold Spring Harbor, New York 2001, ISBN 0-87969-577-3.

8. Osoegawa K, de Jong PJ, Frengen E, Ioannou PA (May 2001). "Construction of bacterial artificial chromosome (BAC/PAC) libraries". *Curr Protoc Hum Genet*. Chapter 5: Unit 5.15. doi:10.1002/0471142905. hg0515s21. PMID 18428289.

9. Hartwell, Leland (2008). *Genetics: from genes to genomes*. Boston: McGraw-Hill Higher Education.ISBN 0-07-284846-4.

10. Kim, H. R., Yang, T. J., Kudrna, D. A. and Wing, R. A. 2004. Construction and Application of Genomic DNA Libraries. *Handbook of Plant Biotechnology*. 2:1:5. DOI: 10.1002/0470869143.kc005

11. Schwartz DC, Cantor CR (1984). "Separation of yeast chromosome-sized DNAs by pulsed field gradient gel electrophoresis". *Cell*. 37 (1): 67–75. doi:10.1016/0092-8674(84)90301-5. PMID 6373014.

12. Shizuya H, Birren B, Kim UJ, Mancino V, Slepak T, Tachiiri Y, Simon M. Cloning and stable maintenance of 300-kilobase-pair fragments of human DNA in Escherichia coli using an F-factor-based vector. *Proc Natl Acad Sci U S A*. 1992 Sep 15;89(18):8794-8797.

13. Hari Prasad O, Nanda Kumar Y, Reddy OV, Chaudhary A, **Sarma PV**. Cloning, expression, purification and characterization of UMP kinase from *Staphylococcus aureus. Protein J*. 2012;31(4):345-352.

14. Prasad UV, Vasu D, Kumar YN, Kumar PS, Yeswanth S, Swarupa V, Phaneendra BV, Chaudhary A, **Sarma PV**. Cloning, Expression and Characterization of NADP-Dependent Isocitrate Dehydrogenase from *Staphylococcus aureus. Appl Biochem Biotechnol*. 2013;169(3):862-869.

15. Prasad UV, Vasu D, Yeswanth S, Swarupa V, Sunitha MM, Choudharry A, **Sarma PVGK**. Phosphorylation controls the functioning of *Staphylococcus aureus* Isocitrate dehydrogenase - favours biofilm formation. *Journal of Enzyme Inhibition and Medicinal Chemistry*. 2015;30(4):655-661.

16. Swarupa V, Chaudhury A, Krishna Sarma PV. Effect of 4-methoxy 1-methyl 2-oxopyridine 3-carbamide on Staphylococcus aureus by inhibiting UDP-MurNAc-pentapeptide, peptidyl deformylase and uridine monophosphate kinase. *J Appl Microbiol*. 2017;122(3):663-675.

Chapter 2

CLONING OF DNA FRAGMENTS INTO PHAGE VECTORS

BACKGROUND

Bacteriophages are viruses that infect host bacteria, propagate in the host and come out of the bacteria either by lysing the host or thrown out as intact phage particles. These bacteriophages are of three types a) Transducing phages which are able to transfer bacterial genes and take them from one bacterial cell to another through the process of transduction. b) Filamentous phage is a kind of virus of bacteria, known by its filament-like or rod-like shape. Filamentous phages, usually contain a genome of single-stranded DNA and infect Gram-negative bacteria. c) Temperate phages, the ability of some bacteriophages (for example λ phage) to display a lysogenic life cycle where phage genome integrates into host bacterium's chromosome, together becoming a lysogen as the phage genome becomes a prophage, also, able to grow and lyse the host bacteria.

Phages infecting *E. coli* are highly exploited and used as vectors across the globe because the sequences of these phages were completely deciphered, easy to handle and small in size compared to phage identified and isolated from other bacteria like *Streptomyces* and *Mycobacteria*. Transducing phages are T phages from T3 to T7 that infect *E. coli* and the phage products such as enzymes, proteins, etc., have been purified and characterized. These molecules are routinely used in various genetic manipulations and in the probe development. Two phages of *E. coli* are largely used as vectors for cloning and creating genomic and cDNA libraries; they are M13 and λ phages.

M13 PHAGE

This phage was first discovered in the Munich city (Germany) (Hofschneider 1963) and M in M13 represents for that city. This phage infects *E. coli* which has F plasmid; the conjugation tube formed from the tra genes of F plasmid is the site for M13 phage infection. This phage has single stranded (SS) circular genome of 6409 bp. There are two more similar phages which were discovered named as f and fd phages which differ only by three bases (6407bp). Upon entry into the *E. coli* host, the phage (+) gets converted into double stranded form which is known as replicative form (RF 1) and phage in this form is exploited for genetic manipulation. These phages upon growing in *E. coli* slow down the host bacterial growth by about 2.5 folds, and plaques are screened as translucent zone over opaque *E. coli* culture in LB agar plates.

M13 phage vectors have been developed by J. Messing and J. Vriera (1978). They have successfully cloned 450bp of *lac Z* gene which encodes for α-peptide of β-galactosidase (*lac Z*). In the NH_2 terminal of this gene multiple cloning sites were introduced and with strong terminator, these vectors were named as M13mp. This α-peptide complements the β-galcatosidase post-translationally in the host to form active β-galactosidase enzyme. This enzyme can degrade a chromogenic compound X-gal in the presence of IPTG to give blue plaques, this serves as indicator for identifying the recombinants and non-recombinants upon cloning the desired DNA fragment in the multiple cloning site. The host *E. coli* strains contain the rest of the lac operon, i.e., lac Z gene ▲ M13, lac A, and lac Y in the F plasmid. The lac inducer and the sequence encoding the α-peptide of β-galactosidase gene are present in the M13 phage vector. Development of these vectors has revolutionized the cloning, sequencing and *in vitro* mutagenesis strategies as the desired DNA fragment is obtained in single stranded form and can serve as template to perform all the studies.

M13 phage encodes 10 genes, of them gene III (4 to 5 molecules) and gene VIII (~2700 molecules) products forms the tail and capsid of the phage. Gene V product is essential for converting the RF to SS state where it forms into proper phage and is thrown out of the host *E. coli* into the culture filtrate. In this SS state, the DNA is highly stable as it is resistant to most of the endo and exonucleases. The gene III is usually replaced with our own gene of interest which will be directly expressed in tail region and can be directly used in the identification of the desired gene of interest—this is called as phage display technique. M13 mp18 and M13 mp19 vectors are largely used in the cloning of several genes, these vectors contain multiple cloning sites which are present exactly in the opposite orientations and the resultant proteins are usually fusion proteins fused with α-peptide of β-galactosidase.

Materials

A. Media

1. Luria Bertani (LB) broth: 10g Bacto tryptone, 5g Yeast extract and 5g NaCl dissolve in 1 litre and adjust the pH to 7 with 10 N NaOH

2. LB agar: Add 1.5 g agar to 100ml of LB broth and autoclave at 121 °C or 15lbs pressure for 20 minutes. Pour the LB agar on 100mm sterile petri plate in Laminar Air Flow to ensure no air bubble is present and allow the agar to solidify in the sterile condition. Keep the plate in 37 °C for 30 minutes to dry the moisture.

3. Top agar: Add 0.7g agar to 100ml of LB broth and autoclave at 121 °C or 15lbs pressure for 20 minutes.

4. SOB Medium: 10g Bacto tryptone, 5g Yeast extract, 0.5g NaCl, 10ml of 250mM KCl and adjust the pH to 7 with 10 N NaOH.

5. 2M Glucose solution: Dissolve 36 g of glucose in 50 ml distilled water and make up the volume to 100 ml with distilled water. Sterilize by Millipore filtration.

6. 2M $MgCl_2$: This solution is made by dissolving 19 g of $MgCl_2$ in 70 ml of distilled water and sterilize by Millipore filtration.

7. SOC medium: 200ml of SOB medium containing 2ml of 2M Glucose and 2M $MgCl_2$.

B. Buffers for Transformation

a) 1M $CaCl_2$. $2H_2O$ or Transformation buffer (TB): MES [2-(N-morpholino)ethanesulfonic acid] buffer contains: 10mM MES pH 6.3, 45mM $MnCl_2.4H_2O$, 10mM $CaCl_2.2H_2O$, 100mM KCl, 3mM Hexaminecobalt chloride, prepare in autoclaved distilled water. After all the salts are dissolved sterilize by Millipore filtration (filters anything between 0.45µM to 0.2µM pore size).

C.   Buffers for the isolation of M13 double strand and single strand phage DNA:

1.   TEG Buffer

   20% Glucose

   50mM Tris-HCl

   50mM EDTA.   Adjust the pH to 8.0

2.   Lysis Buffer

   1% SDS

   0.2N NaOH

   Note: For preparing 10 ml of lysis buffer. Take 8.8 ml of sterile distilled water and to that add 0.2 ml of 10N NaOH and 1 ml of 10% SDS. Do not add SDS first as it will precipitate when NaOH is added to the solution.

3.   M Potassium acetate pH 4.8

   Dissolve 29.44 g potassium acetate in 25 ml of distilled water and adjust the pH to 4.8 with glacial acetic acid. Finally make up the volume to 100 ml of distilled water and sterilize by autoclaving at 15 lb / sq.in for 15 minutes.

4.   RNase. Dissolve 10 mg / ml pancreatic RNase in 50mM Tris-HCl pH 7.5 containing 0.15M NaCl. Keep the above solution in boiling water bath for 10 minutes, this is to destroy the DNase function.

5.   Solution B: Chloroform: Isoamyl alcohol 24:1 (v/v)

6.   Isopropanol (ice cold)

   Chill the isopropanol by keeping it in ice before using.

7.   70% Ethanol

   Mix v/v 70 ml of 100% ethanol and 30 ml of sterile distilled water.

8.   10 mg / ml Lysozyme solution.

Dissolve 10 mg of lysozyme solution in 1 ml of TEG buffer and leave it on ice.

9. TE Buffer

    10mM Tris-HCl

    1mM EDTA, pH 8.0

10. Polyethylene glycol (PEG) buffer: 20% PEG6000 dissolved in 2.5M NaCl solution.

Method

Transfection of M13 double stranded DNA into E. coli JM109 or *E. coli* TG1:

1. Streak the *E. coli* JM109 or *E. coli* TG1 culture on LB agar plate and incubate at 37 ºC for overnight and pick single colony and inoculate in 5ml LB broth and grow up to mid-log phase at 37 ºC in an environmental shaker with constant shaking at 120 rpm.

2. Inoculate 1 ml of this growing culture in 200ml SOC medium and allow the culture to grow up to mid log phase (O.D=0.5) and keep the cells in ice for 30 minutes.

3. Centrifuge at 7,000 rpm for 10 minutes at 4 ^{0}C, discard the supernatant and suspend the pellet in 20ml of ice cold 0.1M $CaCl_2.2H_2O$ or TB and keep it in ice for 30 minutes.

4. Centrifuge at 7,000 rpm for 10 minutes at 4 ºC, discard the supernatant and suspend the pellet in 2ml of ice cold 0.1M $CaCl_2.2H_2O$ or TB and keep it in ice.

5. Take 200µl of cells in a sterile 1.5ml microcentrifuge tube and add 10-50pmoles of M13ds DNA and keep it in ice for 30 minutes.

6. Keep the tube of step 5 at 42^0 C for 90 seconds.

7. Add 200 µl of *E. coli* JM109 or TG1 culture in sterile 5ml glass tube. Add 200 µl of step 6 tube and 3 µl of 1M IPTG and 30 µl of 20mg/ml X-Gal (20mg of X-Gal = 5'Bromo 4'Chloro 3' indolyl β-D galactoside dissolved in N' dimethylformamide) and add 2.66 ml of molten top agar maintained at 50ºC and quickly pour on a LB agar plate and allow the agar to solidify and keep the plate at 37^0 C for 8-10 hours and observe light blue coloured plaques in the plate.

8. Grow *E. coli* JM 109 or *E. coli* TG1 in 5-10ml of LB broth at 37 ^{0}C with a constant shaking at 150-200 rpm up to late log phase (OD$_{540}$= 0.5) and take sterile tooth prick and punch only one plaque and quickly put in the tube and allow it grow for further 6-8hours.

9. Take 1.5 ml of the above grown culture in a sterile 1.5ml microcentrifuge tube and centrifuge the cells at 7,000 rpm for 90 seconds at room temperature in a microcentrifuge.

10. Retain the supernatant for single strand M13 DNA isolation and add 0.5 ml of TEG buffer and suspend the pellet and centrifuge the cells at 7,000 rpm for 90 seconds at room temperature in a microcentrifuge.

11. Discard the supernatant and suspend the pellet in 150μl of ice cold lysozyme solution and leave it in room temperature for 10 minutes.

12. To the same above mixture add 50 μl of 10mg/ml RNase and leave at 37^0 C for 10 minutes.

13. Now add 300 μl of lysis buffer and mix it carefully such that clear transparent solution is formed and leave it on ice for 10 minutes.

14. Now add 250 μl of 3M Potassium acetate solution pH 4.8 slowly and leave it on ice for 30 minutes.

15. Centrifuge at 10,000 rpm for 10 minutes at 40^0 C, discard the pellet and to the supernatant add 1:1 (v / v) of ice cold isopropanol mix by vortexing and leave it on ice for 2-3 minutes and centrifuge at 10,000 rpm for 10 minutes at 4^0 C.

16. Discard the supernatant and add 0.5 ml of 70% ethanol and centrifuge at 10,000 rpm for 10 minutes at 4 ^{0}C. Discard the supernatant and air dry the pellet. Dissolve the pellet in 50μl of TE buffer. Analyze the plasmid DNA by running one percent agarose gel.

Single strand M13 DNA Isolation

1. The supernatant obtained in step 10 is used to isolate the single stranded M13 phage DNA. The supernatant is distributed equally in two autoclaved 1.5ml microcentrifuge tubes, to that add 100μl of PEG buffer and incubate at 37 ^{0}C for 30minutes to 1hour.

2. Centrifuge the mixture at 10,000rpm for 15 minutes at 4 ^{0}C. Discard the supernatant and suspend the pellet in 100 to 200 μl of TE buffer.

[Note: Here pellet appears as round shape attached to one side of the micro-centrifuge tube.]

3. Discard the supernatant and dissolve the pellet in 100µl of TE buffer.

4. Add equal volume of solution B and vortex briefly for 3 to 5 minutes at room temperature. Repeat centrifugation step.

5. Carefully pipette out the supernatant.

6. Air dry the pellet and dissolve the pellet in 50 µL of TE buffer at room temperature. This can be used in the recombinant DNA experiments such as in dideoxy method of DNA sequencing or site directed muta-genesis.

References

1. Ray DS, Dueber J, Suggs S. Replication of bacteriophage M13 IX. Requirement of the *Escherichia coli* dnaG function for M13 duplex DNA replication. *J. Virol.* 1975;16(2):348-355.

2. Messing J. New Ml3 vectors for cloning. *Meth. Enzymol.*1983; 101:20-79.

3. Sanger F, Nwklen S, Coulson A R. DNA sequencing with chain terminating inhibitors. *Proc. Natl. Acad. Sci.* USA. 1977;74:5463-5467.

4. Messing J. (1993) Ml3 clonrng vehicles, m DNA Sequencing Proto-cols, *Methods in Molecular Biology*, vol. 23 (Griffin, H. G. and Griffin, A. M., eds.), Humana, Totowa, *NJ*, pp. 9-22.

5. Yarnsch-Perron C, Vieira J, Messing J. Improved Ml3 phage cloning vectors and host strains nucleotide sequenceso f the M13mp18 and pUC 19 vectors. *Gene.* 1985; 33:103-119.

6. Sambrook J, Russell DW, *Molecular cloning.* A laboratory manual, Cold Spring Harbor Laboratory Press, Cold Spring Harbor, New York 2001.

7. Adams RCP, Burdon RH, Campbell AM, Leader DP and Smellie RMS, *The Biochemistry of the Nucleic acids* 9th edition, Chapman and Hall London, 1981.

LAMBDA PHAGE

E. Lederberg (1950) discovered this *E. coli* infecting virus, and he found this phage genome either gets integrated with host chromosome and stays in prophage condition known as lysogeny or replicates the phage genome to many copies and packs them as phage particles before killing the bacteria and coming out in the form of plaques which is known as lytic cycle. This lambda phage consists of head (capsid) and tail fibres as the protein part which encloses a linear double stranded DNA having 12 base pair cohesive ends known as cos sites. The molecular size of phage DNA is 48,502bp. Bacteriophage Lambda attaches to the *E. coli* through its J protein of the tail tip. This J protein binds with the maltose outer membrane porin (the product of the lamB gene) of *E. coli*, a part of the maltose operon. The linear phage genome is injected into the bacterium through the outer membrane. The DNA passes through the mannose permease complex in the inner membrane (encoded by the man XYZ genes) and immediately circularises using the cos sites (**Cos site Left GGGCGGCGACCT and Cos site Right CCCGCCGCTGGA**), 12 base GC rich cohesive "sticky ends" and this reaction is catalyzed by host DNA ligase. On entering into *E. coli,* the phage DNA circularizes at the ends and forms covalently closed circular form, where it replicates largely by θ shaped replication mode; however, expression of either gam genes from phage genome or recBCD genes of host *E. coli* converts the phage replication to rolling circle model. In this mode of replication the concatemers are packaged into phage particles (Figure 2).

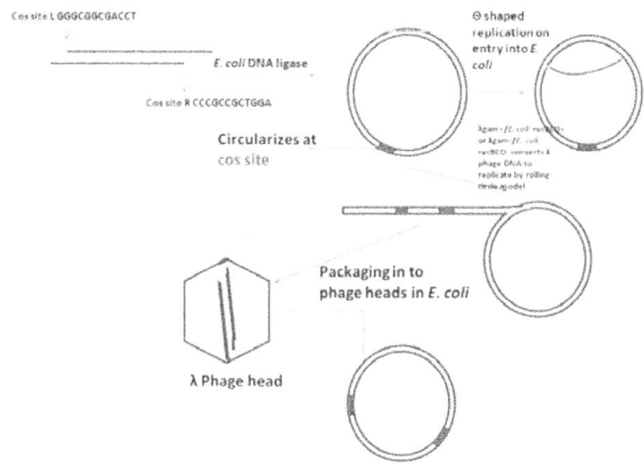

Figure 2   Replication of lambda phage in *E. coli* host

In the host organism, negative supercoils in the circular chromosome are created by host DNA gyrase causing AT rich regions to relax and initiates transcription which starts from the constitutive P_L, P_R and $P_{R'}$ promoters producing the 'immediate early' transcripts. At first, N and cro genes are transcribed and translated producing N, Cro and a short inactive protein.

i.   Cro binds to OR_3 operator in the P_R promoter, preventing access to the P_{RM} promoter, averting the expression of cI gene. The anti-terminator N binds to the two Nut (N utilisation) sites; one in the N gene in the P_L reading frame, and one in the cro gene in the P_R reading frame.

ii.   When RNA polymerase transcribes these regions, it employs the N protein which forms a complex with several host Nus proteins. This complex skips through most termination sequences, as a result the 'late early' transcripts which includes the N and cro genes along with cII and cIII genes, and xis, int, O, P and Q genes are transcribed immediately.

iii.   The cIII protein stabilizes the formation of cII protein and guards from proteolysis by FtsH (a membrane bound essential *E. coli* protease) by acting as a competitive inhibitor. This inhibition induces a bacteriostatic state, which favours lysogeny. In this state constant expression of cI gene maintains the lysogenic state of the phage genome. This cI protein then binds to OR_1 and OR_2 preventing access to the P_R curtailing the expression of cro gene. Thus, the protein concentrations of cro and cI in the host dictate whether the phage enters into lytic cycle or lysogenic state.

iv. The phage DNA once it enters into lytic cycle all the proteins of head and tail genes are expressed and the DNA gets linearized at the cos site by A gene product and it is then packaged into phage head and tail. The S, R and Rz proteins lyse the host *E. coli* and release the phages. For phage DNA to be successfully packaged into phage head, minimum of 75% (36.37Kbp) of the size of phage DNA must be present and maximum size of the DNA that can be accommodated into phage head is 105% (50.9 Kbp) of the phage DNA. For packaging the phage DNA, only cos site and A gene product are essential requirements while head and tail proteins can be arranged into phage heads in the test tubes using E and D gene products of lambda phage (*in vitro* packaging). These properties are aptly exploited in the construction of λ phage vectors.

LAMBDA PHAGE VECTORS

The Lambda phage vectors are constructed with following broad objectives: (i) the presence of cloning sites only in the dispensable fragments, (ii) the capacity to accommodate foreign DNA fragments of various sizes, (iii) the presence of multiple cloning sites, (iv) an indication of incorporation of DNA fragments by a change in the plaque type, (v) the ability to control transcription of a cloned fragment from promoters on the vector, (vi) the possibility of growing vectors and clones to high yield, (vii) easy and ready recovery of cloned DNA, and (viii) introduction of features contributing to better biological containment. To fulfil all these objectives in the lambda phages, essential regions and non-essential regions have been identified. As it is clear that the lysogenic state of the λ phage does not support the above said objectives, therefore it is referred as non-essential region, whereas the lytic cycle is known as essential region.

There are two types of vectors that can be constructed using λ phage DNA they are (a) insertional vectors and (b) replacement vectors. Insertional vector is type of vector in which λ phage DNA is digested with a RE 1 which has single site in the DNA and new foreign DNA is inserted in the RE 1 site (Figure 3). Replacement vector is a type of vector in which λ phage DNA is digested with two different restriction enzymes or single restriction enzyme having two sites in the λ phage genome and is replaced with foreign DNA (Figure 3). These recombinant phages are propagated in the *E. coli* host which has the following characteristic features:

a.   Bacteriophage λ enters *E. coli* host as a linear molecule that rapidly circularizes at the cos site and, in the initial phase of infection, the phage genome replicates by a bidirectional θ- type mechanism, giving monomeric circles. Thus, replication converts to a rolling-circle σ-type mechanism due to the presence of gam gene expression from the phage genome, creating linear concatemers that are suitable substrates for packaging into phage heads (Figure 2). Such host *E. coli* is a preferred strain for propagating λ vectors.

b.   Rolling-circle replication is inhibited by host *E. coli* RecBCD, which readily degrades the linear concatemeric DNA. Thus, for efficient propagation by rolling-circle replication requires a recBC– sbcB– or recD– host. Alternatively, the exonucleolytic activity of RecBCD can be inhibited by the λ gam gene product, which may be carried on the λ vector itself or on a separate plasmid inside the host. Such *E. coli* strains are mostly favoured for the growth of λ vectors.

c.   In recBCD+ strains of *E. coli*, λ vectors with gam will result in the production of the λ phage progeny to multimeric circles that are the substrates for packaging catalyzed by either λ-encoded Red recombinase or host RecA. Most λ vectors are gam– red– and, therefore, require a RecA+ *E. coli* host for propagation.

d.   The occurrence of an octameric sequence GCTGGTGG, termed a χ (chi) site, in the gam– λ genome are useful in the replication in recBC+ host. The χ site in the λ recombinant phages results in the increased recombination, by a RecBCD dependent pathway. This requires RecA, causing more efficient conversion of monomeric to multimeric circular forms. It is very certain that cloned sequences containing χ site will be overrepresented in genomic libraries prepared in gam– χ– of λ vectors if grown in a recBC+ host. Therefore adequate expression of a gene cloned in a λ vector requires proper *E. coli* host for its survival and obtaining the desired gene product.

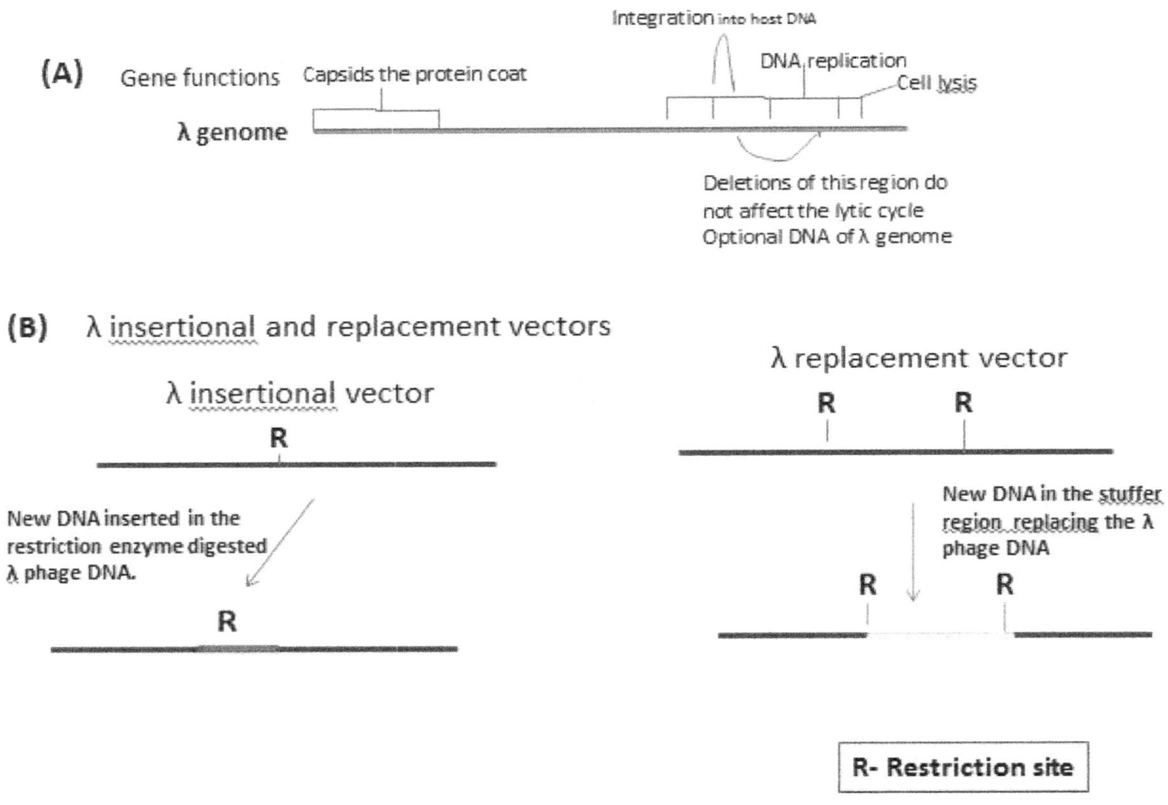

Figure 3 Schematic representation of development of lambda phage vectors

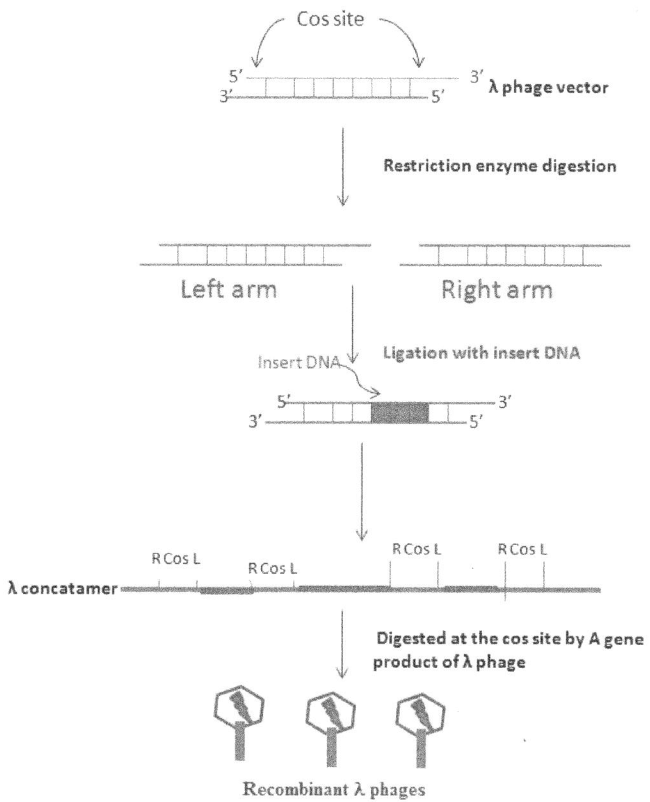

Figure 4   Cloning in lambda phage vectors

Transfection of lambda phage into *E. coli* **Y1090**

Materials

1. Luria Bertani (LB) broth: Dissolve 5 g Yeast extract, 10g of peptone and 5g of NaCl in 750ml distilled water and adjust the pH to 7 with 10N NaOH. Finally make up the media to 1000ml with distilled water. Sterilize by autoclaving at 15lb/sq or 121 °C for 20 minutes.

2. Top agar (0.7% LB agar): Dissolve 0.7g agar in 100 ml of LB broth and Sterilize by autoclaving at 15lb/sq or 121 °C for 20 minutes.

3. LB agar: Add 1.5grams of bacteriological agar to 100ml of LB broth. Sterilize by autoclaving at 15lb/sq or 121 °C for 20 minutes.

4. 20% Maltose: Dissolve 20g of maltose in 100ml distilled water; sterilize by 0.22 micron Millipore filtration.

5. SM Buffer: This buffer is used for storage and dilution of bacteriophage λ stocks. Per liter

NaCl	5.8g
$MgSO_4.7H_2O$	2.0g
1M Tris-HCl (pH 7.5)	50ml
2% gelatin solution	5ml

Dissolve all the above materials in 1000 ml of distilled water and sterilize by autoclaving at 15lb/sq or 121 °C for 20 minutes. Add 2g of gelatin in 100ml distilled water and sterilize by 0.22 micron Millipore filtration.

6. TM Buffer: 1M Tris-HCl pH 7.5   50ml and 2g of $MgSO_4.7H_2O$, dissolve in 1litre of distilled water and sterilize by sterilize by 0.22 micron Millipore filtration.

7. TE Buffer: 10mMTris-HCl and 1mM EDTA pH 8 sterilize by autoclaving at 15lb/sq or 121 °C for 20 minutes.

8. Solution A.: *Phenol saturated with 1M Tris-HCl pH 8 and stored in 0.1M Tris-HCl

9. Solution B: Chloroform: iso-amyl alcohol: 24: 1 (v/v) mixture, respectively

10. Solution C: Mix solution A and solution B in 1:1 v/v ratio and is used for the extraction of the DNA

11. Ice cold isopropanol

12. 70% Ethanol: 70 ml of 100% ethanol and 30 ml of sterile distilled water are mixed in a sterile 100ml flask

Method

13. Streak the host *E. coli* strain on LB agar plate and incubate overnight at 37 °C. Pick single colony and inoculate in 250ml LB broth that contain 2% maltose and allow the bacteria to grow up to OD_{540}=0.9-1 (which has approximately 1x 10^9 bacteria/ml).

a. Centrifuge the bacterial culture at 7,000 rpm for 5 minutes at 4 ^{0}C, discard the supernatant and suspend the pellet in 1ml of TM buffer.

b. Take 0.1ml of above culture and mix 1:100 diluted bacteriophage λ stock in SM buffer and incubate at 4 ^{0}C for 30minutes.

c. Add 3ml of molten top agar to the above mixture (4) and pour immediately on top of a LB agar plate and allow the plate to solidify.

d. Incubate the plate at 37 ^{0}C for 10-12hours. Identify the formations of plaques.

e. Grow the host *E. coli* strain up to mid log phase OD540=0.45-0.55 in LB broth containing 2% maltose. From the above plate pick a plaque with the help of sterile tooth prick and mix with the growing culture and incubate at 37 ^{0}C for 6-7h with constant agitation.

f. After the incubation centrifuge at 10,000 rpm for 10 minutes at 4 ^{0}C. The supernatant contains the bacteriophage lysate store at 25% glycerol (25% glycerol stock: mix 750μl of lysate and 250μl of 100% sterile glycerol in a 1.5ml microcentrifuge tube) and extract the DNA from the lysate.

g. Mix the above lysate with equal volume of solution C and vortex for 2-3minutes and centrifuge at 10,000 rpm for 10 minutes at 4 ^{0}C.

h. Take out cleanly the aqueous phase in a sterile 1.5ml microcentrifuge tube and add equal volume of ice cold isopropanol and centrifuge at 10,000 rpm for 10 minutes at 4 ^{0}C. Discard the supernatant and suspend the pellet in 250μl of 70% ethanol and centrifuge at 10,000 rpm for 10 minutes at 4 ^{0}C. Discard the supernatant and air dry the pellet and suspend in 50μl of TE buffer.

i. Analyse the DNA in 0.7-1% agarose gel electrophoresis.

GENOMIC LIBRARY CONSTRUCTION

In a genomic library the total genomic DNA from a single organism is stored in a population of identical vectors, each containing a different insert of DNA present in an organism like *E. coli*. In order to construct a genomic library, the organism's DNA is extracted from cells or bacteria and then digested with a restriction enzyme to cut the DNA into fragments of a specific size. The fragments are then inserted into the vector using DNA ligase. This is followed by the transformation of ligation mixture (Ligation mixture means vector DNA + insert DNA) into a host organism - commonly a population of *Escherichia coli*. The colonies thus obtained represent with each bacterium containing one vector molecule containing insert DNA. This allows easy amplification and retrieval of specific clones from the library for analysis. Lambda phage vectors, both insertional or replacement can be successfully employed in the construction of the library. Using gene specific probes, the desired gene can be successfully identified by plaque hybridization technique (Figures 4 and 5).

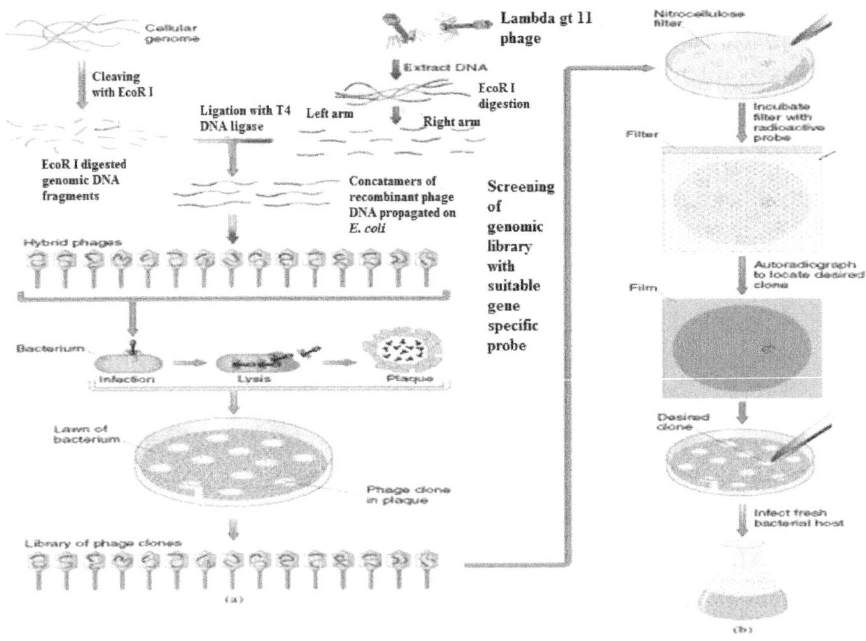

Figure 5   Genomic library preparation in lambda phage vectors and screening by plaque hybridization

COMPLEMENTARY DNA SYNTHESIS AND CDNA LIBRARY CONSTRUCTION

Total mRNA isolation from the tissue

Eukaryotic mRNA is characterized by having poly A tail at the 3' end which is exploited in the purification of mRNA from the total RNA (earlier section 3.1). Approximately 1% of the total RNA are mRNA present in all actively growing cells and shelf lives of these molecules is very less and are rapidly degraded in the cells, however, there are mRNA's and hnRNA's which are found to be present for years in the cells but they are very rare and are present in very low levels. These mRNA molecules from total RNA preparation can be isolated by passing through an affinity column made up of cellulose/Sepharose which is tagged with oligo dT. When total RNA is passed through the column, all rRNA's and tRNA's comes out in the flow through, while mRNA's gets attached with oligo dT through hydrogen bonding and stays in the column. When the buffer is having SDS, the hydrogen between poly A of mRNA and oligo dT are broken and pure mRNA is eluted out. This pure mRNA is used in the construction of cDNA libraries, in the reverse transcript polymerase chain reaction (RT-PCR) and in real time PCR to assess the levels of mRNA present in the cell. The percent isolated mRNA levels are calculated as:

Concentration of mRNA / concentration of total RNA*100

cDNA Construction

Once mRNA is purified, oligo-dT (a short sequence of deoxy-thymidine nucleotides) is tagged as a complementary primer which binds to the poly-A tail providing a free 3'-OH end that can be extended by reverse transcriptase to create the complementary DNA strand. Now, the mRNA is removed by using RNAse enzyme leaving a single stranded cDNA (sscDNA). This sscDNA is converted into a double stranded DNA with the help of DNA polymerase. However, for DNA polymerase to synthesize a complementary strand a free 3'-OH end is needed. This is provided by the sscDNA itself by generating a hairpin loop at the 3' end by coiling on it. The polymerase extends the 3'-OH end and later the loop at 3' end is opened by the scissoring action of S1 nuclease. Restriction endonucleases and DNA ligase are then used to clone the sequences into phages (M13mp18 or λgt 11 vectors). In order to clone into λgt 11 for full length expression of cDNA, the generated cDNA are treated first with EcoR I methylase. This is to block the digestion of internal EcoR I sites and this is followed by ligation with linker DNA having single EcoR I site. Digest this DNA with EcoR I and ligate using T4 DNA ligase. The clones are then selected, commonly through the use of X gal and IPTG. Using gene specific probes the desired cDNA can be picked by plaque hybridization (Figure 6).

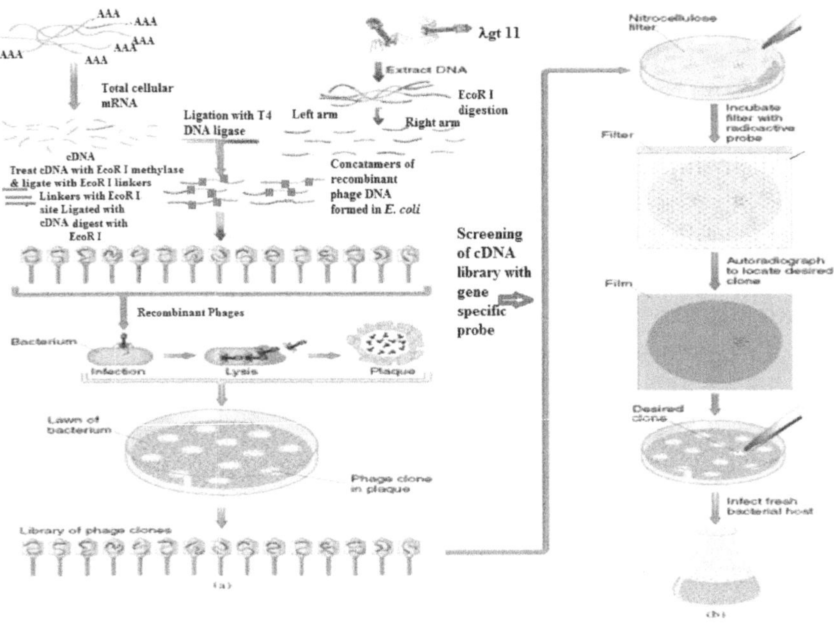

Figure 6   cDNA library preparation using lambda phage vectors and screening by plaque hybridization.

Materials

A. **Restriction endonucleases digestion of Chromosomal DNA**

A.1 Sau3A digestion of chromosomal DNA

10X reaction buffer as supplied by the manufacturer

20 Unit/µl of Sau3A or as defined by the manufacturer

0.5M EDTA, pH 8

50X TAE buffer pH 8.3

Low EEO agarose

Ice cold isopropanol

70% ethanol: 30 ml autoclaved distilled water and 70 ml 100% ethanol.

Chromosomal DNA

Method

1. Take 5µg (25µl) of DNA isolated from the organism in a sterile 1.5ml microcentrifuge tube.

2. Add 50 µl of 10X reaction buffer

3. Add 425 µl of autoclaved distilled water

4. 2µl of Sau3A (20 Unit/µl)

5. Take 5 sterile 1.5ml microcentrifuge tubes and label them as 5', 10' 20' and 40' and add 50µl of 0.5M EDTA.

6. Incubate the above mixture (steps 1 to 4) at 37 ^{0}C exactly after 5' pipette 100µl of the reaction mixture and place it in 5' labelled 1.5 ml microcentrifuge tube and vortex briefly and leave at room temperature.

7. Similarly, pipette out 100µl after 10', 20' and 40' and transfer it to the respective tubes containing EDTA.

8. After 60 minutes to the remaining 100µl add 50µl of 0.5MEDTA vortex and leave at room temperature

9. Perform 1% agarose gel electrophoresis along with Supermix DNA molecular size ladder and undigested DNA in 1X TAE buffer under submerged conditions.

10. Record the gel and cut the desired range of DNA fragments gel and electroelute the fragments in 10mM Tris-HCl, 10mM EDTA pH 8.0 buffer present in the dialysis bag. Perform electroelution for 45 minutes to 60 minutes by keeping the bag at cathode.

11. Take out the buffer containing DNA fragments and add equal volume of ice cold isopropanol and mix the solutions briefly by vortexing following this centrifuge the mixture at 10, 000 rpm for 10 minutes at 4 ^{0}C.

12. Discard the supernatant slowly and add 500µl of 70% ethanol vortex briefly and perform the centrifugation as mentioned in step 11.

13. Discard the supernatant very slowly and air dry the pellet

14. Once all the alcohol is evaporated suspend the pellet in 100µl of TE buffer and use it for the construction of genomic library.

A.2 Restriction endonuclease digestion of λgt11 or M13mp18 vector

1. 10X reaction buffer as supplied by the manufacturer

2. 20 Unit/µl of Bam HI or as defined by the manufacturer

3. 20 Unit/µl Sma I or as defined by the manufacturer

4. 20 Unit/µl EcoRI or as defined by the manufacturer

5. 0.5M EDTA, pH 8

6. 50X TAE buffer pH 8.3

7. Low EEO agarose

8. Ice cold isopropanol

9. 70% ethanol: 30 ml autoclaved distilled water and 70 ml 100% ethanol

10. λgt11 or M13mp18 vector DNA 1µg

11. Calf intestine alkaline phophatase (CIAP) (10Units/µl)

12. Solution B: Chloroform:isoamylalcohol mixed in 24:1 (v/v).

Method

1. Take 1µg (10µl) of λgt11 or M13mp18 vector DNA in a sterile 1.5ml microcentrifuge tube.

2. Add 5 µl of 10X reaction buffer

3. Add 84 µl of autoclaved distilled water

4. 1µl of Bam HI (20 Unit/µl) and incubate at 37 ^{0}C for 1 hour and repeat this step two more times.

5. In case of Sma I enzyme digestion: add 1µl of Sma I (20 Unit/µl) and incubate at 25 ^{0}C for 1 hour and repeat this step two more times. (This is for blunt end ligation of cDNA with vector DNA)

6. In case of cDNA library preparation in λgt11: The lambda phage treated with EcoRI (20 Unit/µl) and incubate at 37 ^{0}C for 1 hour and repeat this step two more times.

7. Add 1µl of CIAP (10 Unit/µl) to the above solution and incubate at 37 ^{0}C for 1 hour and repeat this step two more times.

8. Add equal volume of solution B and vortex for 2 minutes at room temperature and perform centrifugation at 10, 000 rpm for 10 minutes at 4 ^{0}C.

9. Take out the clear aqueous phase in a sterile 1.5ml microcentrifuge tube and add equal volume of ice cold isopropanol and vortex briefly. Perform the centrifugation at 10, 000 rpm for 10 minutes at 4 ^{0}C.

10. Discard the supernatant very slowly and add 250 µl of 70% ethanol and vortex for 2 minutes and perform centrifugation at 10, 000 rpm for 10 minutes at 4 ^{0}C.

11. Discard the supernatant very slowly and air dry the pellet.

12. Dissolve the pellet in 100µl of TE buffer (10mMTri-HCl and 1mM EDTA pH 8).

13. In the case of λgt11 we will get left arm and right arm DNA while for M13mp18 vector linear form of M13mp18 similar to pUC18 plasmid DNA.

A.3 Complementary DNA synthesis

i. *Total mRNA isolation*

Materials

1. DEPC Treated distilled water: The distilled water is made RNase free by treating with diethyl pyrocarbonate (DEPC). Add 1 ml of DEPC to 1litre of distilled water and incubate at 37 °C for 120 minutes; this is to allow DEPC to inactivate ribonucleases. Destroy DEPC by autoclaving at 15 lb / sq.in for 20 minutes and use this water for all buffer preparations, electrophoresis, cleaning of glassware and plastic ware and dissolving RNA.

2. Preparation of glass ware and plastic ware: Soak all glassware in chromic acid for overnight and following day rise in large volumes of tap water and soak for a minute in mild detergent and followed by gentle scrubbing wash it with large volumes of running tap water and follow it up with distilled water. Dry the glassware and soak in DEPC treated water for 1h at room temperature and then bake at 180 °C for 3 hrs in hot air oven. Treat all plastic ware with chloroform saturated with distilled water and immediately soak in DEPC treated water for 1 hour and autoclave at 121 °C for 20 minutes.

3. 2X column loading buffer: 40mM Tris-HCl (pH 7.5), 2mM EDTA, 1mM NaCl and

4. 0.2%SDS (w/v)

5. Elution Buffer: 10mM Tris-HCl (pH 7.5), 0.05% SDS and 1mM EDTA

6. Ice cold isopropanol

7. 70% ethanol: 30ml DEPC treated water and 70 ml absolute ethanol

8. 5M NaCl

9. 10N NaOH

Method

1. Suspend 0.5g to 1g of oligo dT cellulose in 0.1N NaOH

2. Pour the column of oligo dT cellulose (0.5ml to 1ml) in a DEPC treated Dispo column or a Pasteur pipette, plugged with glasswool and sterilized by baking for 3h at 180 ^{0}C. Wash the column with column volumes of sterile DEPC treated water.

3. Wash the column with sterile 1X column loading buffer (dilute from 2X stock using sterile DEPC-treated water) until the pH of effluent is< 8, use pH paper for this measurement.

4. Dissolve the total RNA isolated using earlier methods in sterile double distilled water, and heat the solution at 65 ^{0}C for 5 minutes cool the solution at room temperature and quickly add the 1 volume of 1X column loading buffer.

5. Apply the solution of RNA to the column and immediately collect the fractions of 1ml. When all the RNA has entered the column, wash the column with 1 column volume of 1X column loading buffer while continuing to collect the flow through till the A260 reaches to 0.001.

6. To improve further the RNA collected is heated at 65 ^{0}C for 5 minutes, and reapply to the column and once again collect the flow through.

7. Wash the column with 5-10 volumes of 1X column loading buffer till the A260 reaches to 0.001.

8. Elute the poly (A+) RNA from the oligo dT column with 2-3 volumes of sterile RNA elution buffer. Collect the fractions and pool them.

9. Precipitate the mRNA adding equal volume of ice cold isopropanol and keep in -20 ^{0}C for > 1h and centrifuge at 10, 000 rpm for 15 minutes at 4 ^{0}C. Wash the pellet twice with 70% ethanol and air dry the pellet till all the traces of ethanol is evaporated.

10. Measure the concentration of RNA and analyze the mRNA by running 1.3% formaldehyde gel electrophoresis.

ii. cDNA synthesis

Method

1. Mix RNA sample and primer d(T)23VN in two sterile RNase free micro-centrifuge tubes.

2. Total RNA   6 µl (10 ng–1 µg)

    d(T)23VN (50 µM)   2 µl

    Total Volume   8 µl

3. Denature RNA for 5 minutes at 70°C. Spin briefly and put promptly on ice. This step is optional. However, it improves the cDNA yield for long messenger RNAs and GC rich RNA regions.

4. Add the following components to one tube.

    5X MMuLV Reaction Mix 4 µl (dilute to 1x depending on the manufacturer's composition)

    MMuLV RT (reverse transcriptase) Enzyme Mix 2 µl (Units choose as per the manufacturer's preparation)

    Nuclease free H_2O variable depending on the total RNA 6µl

5. To the negative control tube, add the following:

    5X MMuLV Reaction Mix   4 µl

    H_2O 16 µl

    Incubate the 20 µl cDNA synthesis reaction at 42 ^{0}C for one hour.

    If Random Primer Mix is used, an incubation step at 25 ^{0}C for 5 min is recommended before the 42 ^{0}C incubation.

6. Inactivate the enzyme at 80 ^{0}C for 5 minutes (some protocol performs at 65 ^{0}C for 10 minutes). Dilute reaction to 50 µl with 30 µl H_2O for PCR. The cDNA product should be stored at - 20 ^{0}C after the completion of the reaction.

7. In case of cDNA library preparation in λgt11:

    i.   Take 5µl of cDNA add 5µl 10X EcoR I methylase buffer

ii. Add 1µl EcoR I methylase (10Units/µl) and 44µl autoclaved distilled water and incubate at 37 ^{0}C for 1 hour. Repeat this addition for two more times.

iii. Extract the cDNA by adding equal volume of Chloroform:isoamyl-alcohol (24:1v/v) and centrifuging at 10,000 rpm for 10 minutes at 4 ^{0}C.

iv. Take out the aqueous phase in a sterile 1.5ml microcentrifuge tube and add equal volume of ice cold isopropanol and mix gently and centrifuge at10,000 rpm for 10 minutes at 4 ^{0}C.

v. Discard the supernatant gently and add 250µl of 70% ethanol and mix the solution by gentle vortexing and followed by centrifugation at 10,000 rpm for 10 minutes at 4 ^{0}C.

vi. Discard the supernatant and air dry the pellet.

vii. Suspend the pellet in 10 to 15µl of H_2O and store in -20 ^{0}C.

B. Ligation Reaction:

Materials

1. 10X T4DNA ligase buffer with 1mM ATP cohesive end ligation
2. 10X T4DNA ligase buffer with 1mM ATP blunt end ligation
3. 100U/µl T4 DNA ligase
4. Autoclaved distilled water.
5. EcoR I linker DNA (0.1µg)
6. Packgene extract (follow as per manufacturer's instructions)

Method

1. In a sterile 1.5ml microcentrifuge tube add 5µl of Sau 3A digested chromosomal DNA fragments or cDNAs.

2. To the above tube add 10µl of Bam HI digested M13mp18 vector

3. In case of cDNA library: add 10µl of Sma I digested M13mp18 vector DNA

4. Now add 2µl of 10X buffer either 1 or 2 depending upon cohesive end ligation or blunt end ligation

5. Add 2µl of autoclaved distilled water

6. Incubate the reaction mixture at 10 ^{0}C for overnight after adding 1µl of T4 DNA ligase, this we call it as Ligation mixture.

7. Similarly, prepare ligation control without adding the Sau 3A digested chromosomal DNA fragments or cDNA, in place of that add 5µl of autoclaved distilled water and incubate the tube at 10 ^{0}C for overnight.

8. For preparing cDNA library in λgt11 vector:

9. EcoRI linkers are ligated first to the cDNA treated with EcoR I methylase enzyme:

   a.   In a sterile 1.5ml microcentrifuge tube add 5µl of cDNA treated with EcoR I methylase enzyme

   b.   To the above solution add 5µl of EcoR I linkers

   c.   2µl of 10X blunt end ligation buffer

   d.   7µl of autoclaved distilled water and 1µl of T4 DNA ligase and incubate at 10 ^{0}C for overnight.

   e.   Following day extract the mixture by adding equal volume of Chloroform:isoamylalcohol (24:1v/v) and centrifuging at 10,000 rpm for 10 minutes at 4 ^{0}C.

   f.   Take out the aqueous phase in a sterile 1.5ml microcentrifuge tube and add equal volume of ice cold isopropanol and mix gently and centrifuge at10,000 rpm for 10 minutes at 4 ^{0}C.

   g.   Discard the supernatant gently and add 250µl of 70% ethanol and mix the solution by gentle vortexing and followed by centrifugation at 10,000 rpm for 10 minutes at 4 ^{0}C.

   h.   Discard the supernatant and air dry the pellet.

   i.   Suspend the pellet in 15µl of H_2O and store in -20 ^{0}C or use directly for EcoR I digestion

   j.   To the above solution add 5µl of 10X EcoR I reaction buffer

   k.   1µl of (20Units/µl) EcoR I enzyme and 29µl of autoclaved distilled water and incubate at 37 ^{0}C for 1 hour. Repeat by adding 1µl of EcoR I. Perform steps e to h. Suspend the pellet in 5µl of H_2O.

l.   To the above add 2μl of 10X buffer for blunt end ligation

m.   10 μl of EcoR I digested λgt 11 vector DNA (right arm and left arm)

n.   Add 2μl of autoclaved distilled water and after adding 1μl of T4 DNA ligase Incubate the reaction mixture at 10 ^{0}C for overnight, this we call it as Ligation mixture.

o.   Add 5

p.   Perform transfection as described in sections 1 and 2.

q.   Screen the desired gene/cDNA using gene specific probes following plaque hybridization technique.

PLAQUE HYBRIDIZATION

Screening a genomic library/cDNA library prepared in phage vectors (M13mp18 or λgt11) using gene specific nucleic acid probe is called as **Plaque hybridization**. After successful transfection the recombinants are identified as clear plaques as opposed to blue plaques (the non-recombinants). In these test plates the desired clone is screened using appropriate probe (Probe is prepared having complete or partial sequence homology with the desired gene). First, plates are kept for 30 min at 4^{0}C. The plates are then overlaid with same size of nitrocellulose membrane (NCM). Using a sterile needle pierce the agar plate through NCM on the top, bottom, right and left of the plate and label at back of the plate using a marker. Peel the NCM from the plate carefully and incubate the plate at 37^{0}C for overnight, while placing the NCM on pool of NaOH such that colonies on NCM faces upside and clear surface of NCM are kept on the pool. This is to lyse the bacteria and denature DNA (plasmid as well as chromosomal DNA) and allowing the probe to hybridize with complementary DNA sequence in the plaques. Followed by this the NCM is treated with 0.05MTris-HCl buffer of pH 7.5. The NCM is either exposed to X-ray film or developed adding suitable substrate which is dependent on the probe which is either radio-labelled or labelled with biotin. If it is labelled with biotin block the NCM with gelatine or BSA and then treat with streptavidin coupled to horse radish peroxidise enzyme. Following the incubation and washing of the NCM, immerse the NCM in solution containing 0.03% H_2O_2 and 3'3'

Diaminobenzidine tetra hydrochloride. Observe brown coloured plaques on the NCM. Once the positive clones are identified on the X-ray film or on the NCM, keep the plates exactly matching the holes that were created on the NCM or radio spots on the X-ray film. Pick up the positive clone and transfect it into the host growing in LB broth with the selection pressure added in it. Isolate the positive recombinant phage (λgt 11) from the supernatant and extract the DNA. Further confirm the presence of insert either by PCR or by restriction analysis. In case of recombinant phage obtained from M13mp18 extract the double stranded phage DNA from the pellet following alkaline lysis method and confirm the insert either by PCR or by restriction analysis. The supernatant contains the recombinant single stranded phage (M13mp18) which can be isolated as explained under M13 phage section in this chapter. In this way the desired clone either in λgt 11 phage or in M13mp18 can be stored indefinitely as glycerol stocks in -20 ^{0}C (Figure 6).

Materials

1. 100mM Tris-HCl, pH 7.5

2. 100mM NaOH

3. 20% Dextran Sulfate

4. 10% SDS

5. 20 X SSPE

    Pre-Hybridization buffer

    6 X SSPE

    1% SDS

    5% Dextran sulfate

    100mg /ml heterologous DNA.

    Hybridization Buffer

    6 X SSPE

    1% SDS

    5% Dextran sulfate

    5-10 pmloes of the oligonucleotide Biotinylated probe

Method

1. In 30ml volume take 10 pmole of the probe and add 1 U of T4 DNA polynucleotide kinase and incubate at 37 ^{0}C for 15 minutes with 25ml of (1.5 x 10^7 cpm/ml) g[P^{32}ATP] reaction mixture.

2. Use this probe for the hybridization.

3. Take the plates grown for 12-14hours at 37 ^{0}C and leave then at 4 ^{0}C for 15-30minutes.

4. Now cut the nitrocellulose membrane exact size of the petri plate and place the nitrocellulose membrane on top of the plate such that entire plate is covered with the membrane. Now mark the plate with the help of a sterile needle punch through nitrocellulose membrane and agar plate. Mark at the back of plate with a permanent marker.

5. Carefully lift the membrane such that the plaques stick to the membrane.

6. Make a 3ml pool of 50mM NaOH on Saran wrap paper and place the nitrocellulose membrane; such that plaques face upside and the plane side faces the NaOH. Leave it for 3 minutes.

7. Similarly, place the same membrane on a 3 ml pool of 50mM Tris-HCl pH 7.5 for 3 minutes.

8. In case of nitrocellulose paper bake the paper in vaccum for 90 minutes at 90 ^{0}C. This is not required for nitrocellulose membranes.

9. Place the membrane in a hybridization bag and seal the bag after adding prehybridization buffer in such a manner that there are no air bubbles. Incubate the bag at 42 ^{0}C for 3 hrs with constant agitation.

10. Carefully remove the membrane from the bag and add hybridization buffer (contains probe) and seal the bag similarly and incubate at 42 ^{0}C for overnight with constant agitation.

11. Wash the blot as follows:

    2 x 3 minutes in 2 X SSPE / 0.1% SDS

    2 x 3 minutes in 0.2 X SSPE / 0.1% SDS

    These steps are performed in room temperature.

2 x 15 minutes in 0.16 X SSPE / 0.2 % SDS at 65°C.

NOTE: Here the conditions need to be identified with the GC content of both probe and the formed library. Expose the membrane to X-ray film (KODAK) in the proper cassette and keep it at -70 °C or at -86 °C overnight. The following day develop the film in the developing solution (As per the KODAK manufacturer's protocol) or in the automated X-ray film developing instrument.

If the probe is made by adding biotinylated probe then perform hybridization in the following manner.

12. Block the membrane in 0.5% Tween-20 in 0.1M Tris-HCl, pH 7.5+ 0.15 M NaCl

13. Dilute streptavidin-peroxidase conjugate in 0.5% Tween-20 + 0.1M Tris-HCl, pH 7+ 0.15 M NaCl and immerse the nitro cellulose membrane and incubate at 37 °C for 1hr.

14. Wash the blot with 0.5% Tween-20 + 0.1M Tris-HCl, pH 7.+ 0.15 M NaCl for 4 times with constant shaking.

15. Develop the blot by adding 20 ml of developing reagent (6 mg of 3, 3' diaminobenzidine tetrahydrochloride in 10 ml of 0.01 M Tris-HCl pH 8.0 and 30 ml of H_2O_2).

If the probe is tagged with Fluorescent agent such FITC or PE; observe the membrane under fluorescent microscope. Imaging and analysis can be performed using a xenon arc-lamp-based gel imaging system with 590 ± 10 nm excitation and 620 ± 10 nm emission filters and a 10 seconds exposure time using Alexa Fluor 594 dye-labelled DNA probe.

16. Place the LB agar plate on top of the developed membrane such that the holes match exactly with the markings on the plate. Now pick up the positive clone which exactly matches with the spot with the help of a sterile tooth prick and grow it in 5ml LB broth in which already the host bacteria is grown up to mid log phase (OD_{540} = 0.5).

17. Centrifuge at 7000rpm for 10 minutes at 4 °C and isolate the phage DNA as explained in 1.1 (in case of M13mp 18 phage) and in 1.2 (in case of λgt 11 phage) confirm the insert by PCR or by digestion with appropriate restriction endonucleases.

IMMUNOLOGICAL SCREENING

Cloning in λgt 11 (an expression vector) in the EcoR I site will always result in the formation of a fusion with 53 base pairs upstream from lac Z termination codon and the expressed protein is present as fusion protein where the desired protein is combined with C-terminal end of β-galactosidase. The recombinant protein can be triggered to express by inducing with IPTG (isopropyl-β-D-thio-galactopyranoside) and X-gal (5-bromo-4-chloro-3-indolyl-β-D-galactopyranoside) and is present in the plaques formed after transfection into *E. coli* Y1090 strain. This can be easily screened using polyclonal antibodies raised against the native protein. In this vector the size of insert DNA is ~ 7.2Kb and this can be a cDNA/genomic DNA fragment encoding a gene.

Materials

Same as explained in chapter 1 in immunological screening section.

Method

1. The method till picking up of the plaques on to nitrocellulose membrane is same as explained in above section.

2. Carefully lift the membrane such that plaques stick to the membrane.

3. Incubate the NCM in blocking buffer for 1h at 37 ^{0}C and then perform washing (1 x 5mintues with gentle agitation) 6 times with 30 ml of wash buffer.

4. Incubate the NCM in primary antibody at 37 ^{0}C for 1 hr. Wash the NCM 6 times with wash buffer.

5. Add secondary antibody (anti rabbit) conjugated to horse radish peroxidase. Incubate at 37 ^{0}C for 1 hr followed by 6 times washing with wash buffer.

6. Pour off the last wash buffer and add 20 ml of developing reagent. (6 mg of 3, 3' diaminobenzidine tetrahydrochloride in 10 ml of 0.01 M Tris-HCl pH 8.0 and 30 µl of 30% H_2O_2) and observe the spots specific to the expressed antigen in the recombinant clones.

7. Place the LB plate containing the positive plaques and pick up positive clone with the help of a tooth pick and put it in a LB tube where *E. coli* Y0190 is growing at mid log phase (OD_{540} = 0.5) grow for another 12 hours at 37 ^{0}C.

8. Centrifuge the tube at 7,000rpm for 10 minutes at 4 ^{0}C, the supernatant contains the pure recombinant clone which can be preserved in 30% glycerol and stored indefinitely at -20 ^{0}C.

References

1. Hendrix, R W, Roberts, J W, Stahl, F W and Weisberg, R A (editors) (1983) Lambda H, Cold Spring Harbor Laboratory Press, New York.

2. Sambrook J, Russell DW, *Molecular cloning.* A laboratory manual, Cold Spring Harbor Laboratory Press, Cold Spring Harbor, New York 2001.

3. Adams RCP, Burdon RH, Campbell AM, Leader DP and Smellie RMS, *The Biochemistry of the Nucleic acids* 9th edition, Chapman and Hall London, 1981.

4. Jakobi R, Wiemann S, Pyerin W. Filter-supported preparation of lambda phage DNA. *Anal Biochem.* 1988 Nov 15;175(1):196-201.

5. Manfioletti G, Schneider C. A new and fast method for preparing high quality lambda DNA suitable for sequencing. *Nucleic Acids Res.* 1988;16 (7):2873-2884.

6. Guimont C. Lambda Phage DNA: Method for Rapid Preparation and Further Restriction Study. *Biochemical Education.* 1983; 21(2): 103-105.

7. Young RA, Davis RW. Efficient isolation of genes by using antibody probes. *Proc Natl Acad Sci U S A.* 1983;80(5):1194-1198.

8. Young RA, Davis RW. Yeast RNA polymerase II genes: isolation with antibody probes. *Science.* 1983;222(4625):778-782.

9. Rajagopala SV, Casjens S, Uetz P. The protein interaction map of bacteriophage lambda. *BMC Microbiol.* 2011;11:213.

10. Erni B, Zanolari B, Kocher HP. The mannose permease of Escherichia coli consists of three different proteins. Amino acid sequence and function in sugar transport, sugar phosphorylation, and penetration of phage lambda DNA. *J Biol Chem.* 1987 ;262(11):5238-5247.

11. Werts C, Michel V, Hofnung M, Charbit A. Adsorption of bacteriophage lambda on the LamB protein of Escherichia coli K-12: point mutations in gene J of lambda responsible for extended host range. *J Bacteriol.* 1994;176(4):941-947.

12. Kobiler O, Rokney A, Oppenheim AB. Phage lambda CIII: a protease inhibitor regulating the lysis-lysogeny decision. *PLoS One.* 2007;2(4):e363.

13. Santangelo TJ, Artsimovitch I. Termination and antitermination: RNA polymerase runs a stop sign. *Nat Rev Microbiol.* 2011;9(5):319-329.

14. Deighan P, Hochschild A. The bacteriophage lambdaQ anti-terminator protein regulates late gene expression as a stable component of the transcription elongation complex. *Mol Microbiol.* 2007; 63(3):911-920.

15. Groth AC, Calos MP. Phage integrases: biology and applications. *J Mol Biol.* 2004;335(3):667-678.

16. Burz DS, Beckett D, Benson N, Ackers GK. Self-assembly of bacteriophage lambda cI repressor: effects of single-site mutations on the monomer-dimer equilibrium. *Biochemistry.* 1994; 33(28):8399-405.

CONSTRUCTION OF PHAGE ARTIFICIAL CHROMOSOME

Background

P1 is a temperate bacteriophage (phage) that infects *Escherichia coli* and some other bacteria. When undergoing a lysogenic cycle the phage genome exists as an autonomous plasmid, which is maintained at low copy number like F plasmid, in the bacterium. Unlike λ phage, the genome of this phage P1 does not integrate with host genome.

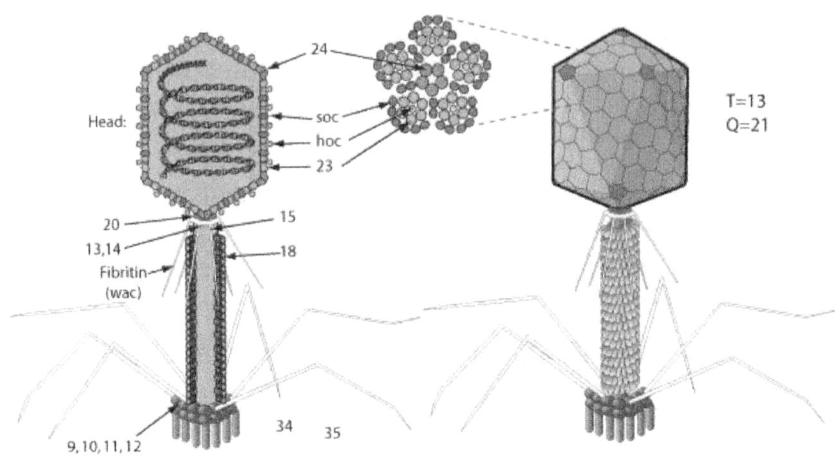

Figure 7   P1 phage structure

The head of P1 phage is an icosahedral structure to which the phage DNA is attached, this preceded by a 220nm long contractile tail and base plate with six tail fibres. The size of the phage head shows an interesting phenomenon; about 80% of population have a head diameter of 85nm, while the rest 20% have a diameter of 65nm. This unique feature is not observed in any temperate phage (Figure 7).

GENOME

The genome of P1 Phage is linear double stranded DNA which is about 94 Kb. The genome is longer, about 120Kbs, than the actual length when in viral particle as it is created by cutting an appropriately sized fragment from a concatemeric DNA chain having multiple copies of the genome. Due to this the ends of the DNA molecule are identical and are referred to as being "terminally redundant". It has large, about 15 kbp, terminal redundancy.

Population of phages exhibit Circular Permutation, using site specific recombination *cre* and *lox P* system. The genome gets circularized and replicates as plasmid molecule inside the host *E. coli*. The host *Escherichia coli* must be recBCD positive for this P1 phage genome to get circularized and exist as covalently closed circular molecule and replicating in rolling circle fashion. The genome contains two origins of replication, oriR which replicates it during the lysogenic cycle and oriL which replicates it during the lytic stage. The genome of P1 codes for 112 proteins and has 5 untranslated genes. It even encodes 3 of its own tRNAs which are expressed in the lytic stage.

Cre recombinase: The Cre protein (encoded by the locus originally named as "Causes recombination", with "Cyclization recombinase" being found in some references) consists of 4 subunits and two domains: The larger carboxyl (C-terminal) domain, and smaller amino (N-terminal) domain. The total protein has 343 amino acids. The C domain is similar in structure to the domain in the integrase family of enzymes isolated from lambda phage. This is also the catalytic site of the enzyme.

loxP site: loxP (locus of X-over P1) is a site on the bacteriophage P1 consisting of 34 bp. The site includes an asymmetric 8 bp sequence, variable except for the middle two bases, in between two sets of palindromic, 13 bp sequences (Figure 8).

A   13bp   8bp   13bp

ATAACTTCGTATA -   NNNTANNN   -TATACGAAGTTAT

The exact sequence is given below; 'N' indicates bases which may vary.

B

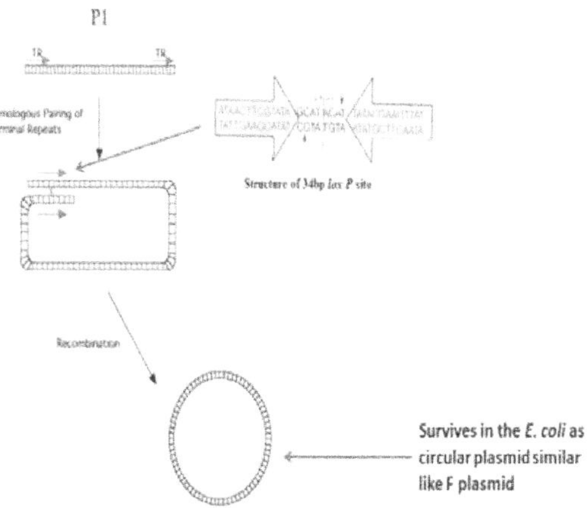

Figure 8   cre-lox P recombination strategy of P1 phage and survival as single copy molecule in host *E. coli.*

When cells that have loxP sites in their genome express Cre, a recombination event can occur between the loxP sites. Cre recombinase proteins bind to the first and last 13 bp regions of a lox site forming a dimer. This dimer then binds to a dimer on another lox site to form a tetramer. Lox sites are directional and the two sites joined by the tetramer are parallel in orientation. The double stranded DNA is cut at both loxP sites by the Cre protein. The strands are then rejoined with DNA ligase in a quick and efficient process. The result of recombination depends on the orientation of the loxP sites. For two lox sites on the same chromosome arm, inverted loxP sites will cause an inversion of the intervening DNA, while a direct repeat of loxP sites will cause a deletion event. If loxP sites are on different chromosomes it is possible for translocation events to be catalysed by Cre induced recombination (Figure 9).

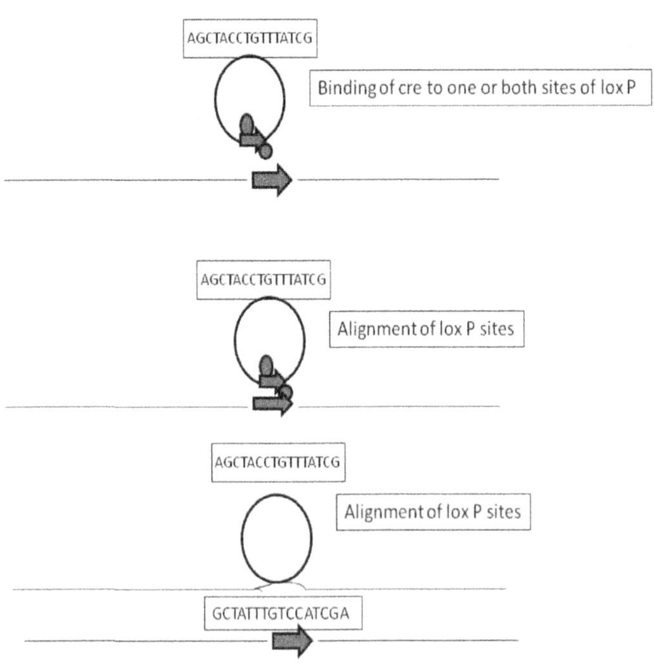

Figure 9   Cre mediated recombination at lox P sites

The bacteriophage P1 cloning was developed as a substitute to YAC and cosmids for cloning high-molecular-weight genomic DNA for two prominent reasons in both the cases—loss of clones due to concatamerization in case of cosmids and due to inherent growth of yeast cells the loss of clones is very high. This is an *in vitro* phage packaging system used to generate >105 clones containing inserts up to <95 kb in size. Introduction of recombinant DNA into the cell utilizing cre and lox P system makes it highly robust, also PAC in the host staying as single copy plasmids without disturbing the *E. coli* like F plasmids makes it convenient to protect the inserted DNA as single copy.

The basic strategies involved in the generation of PAC are as follows:

I.   **Strategy**

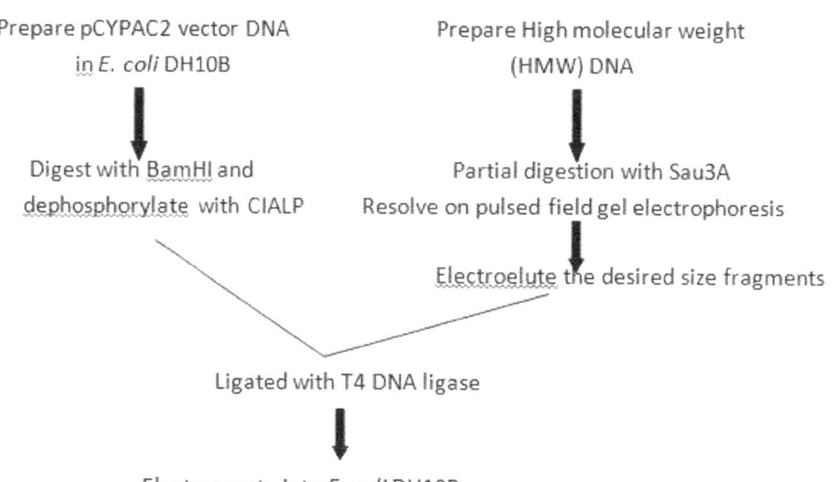

Electroporate into *E. coli* DH10B
Grow the bacteria in LB broth for 1 hour and then plate on LB agar plate containing 20µg/ml kanamycin and sucrose.

Screen the colonies which are negative for sacBII gene, i.e., colonies obtained on LB agar plates containing sucrose

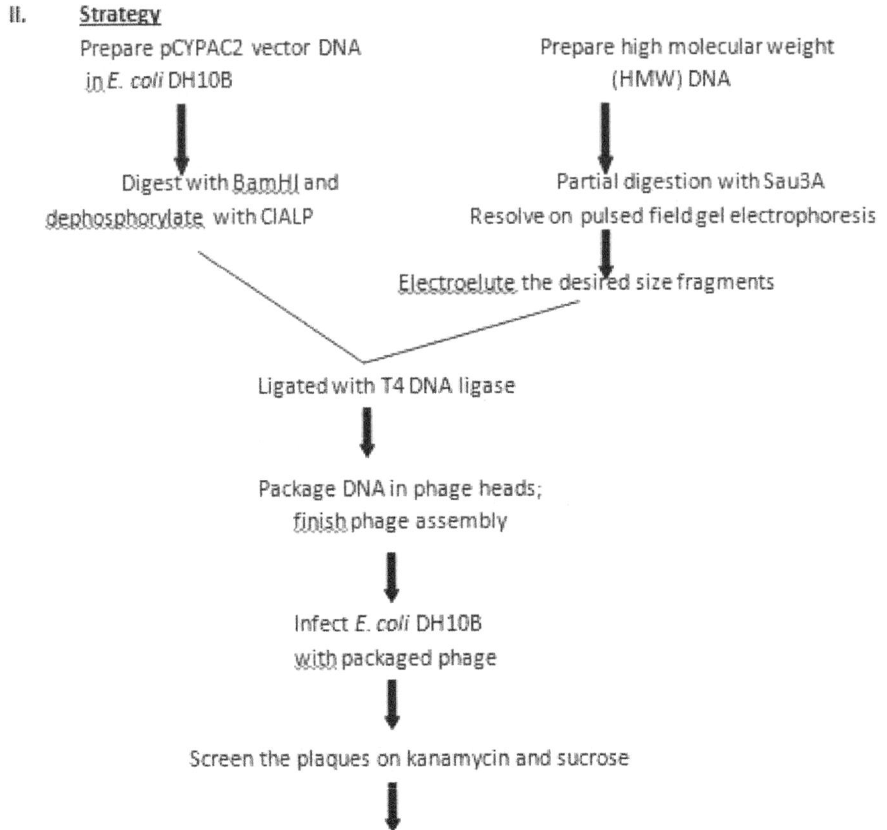

The basic methodology in the preparation of bacteriophage P1 library is digestion of vector in the unique restriction enzyme site followed by ligation with insert DNA, *in vitro* packaging and finally transfecting into host *E. coli* DH10B. The vector used in the P1 system, pCYPAC2, encodes a kanamycin resistance (kan^r) gene to select *E. coli* bacteria containing vector DNA and a sacB gene for positive selection of clones containing insert fragments (Figure 10).

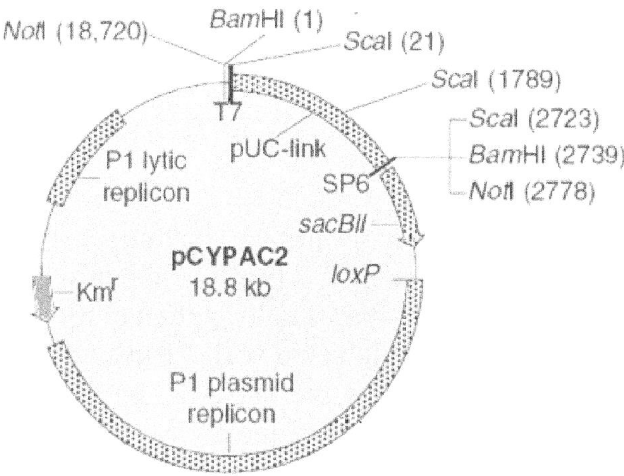

Figure 10   A commercial P1 phage based vector for the preparation of P1 phage artificial chromosome (PAC).

The vector is cleaved at unique ScaI and BamHI restriction sites to generate two "arms" (Figure 10). The BamHI site is situated between the sacB gene and its promoter. Genomic-DNA   generated upon digestion with either Mbo I or with Sau 3A are ligated between these arms at the BamHI-generated ends, disrupting the sacB gene. Recombinant molecules are electroporated into *E. coli* strain containing a constitutively expressed P1 Cre recombinase gene. Cre promotes recombination between P1 loxP recombination sites that flank the cloned insert. The resulting DNA circles contain the cloned insert, the kan[r] gene, a P1 plasmid-replicon-partition system that maintains the DNA as a unit-copy plasmid in the infected cell, and a P1 lytic multicopy replicon regulated by the lac operon. Because cloning of inserts at the BamHI site inactivates sacB, a gene whose product kills bacteria grown in the presence of sucrose, recombinants are selected by growth of transformants on agar plates with kanamycin and sucrose.

IN VITRO PACKAGING

Packaging P1 phage based vectors and P1 phage vectors containing the insert DNA is called as *in vitro* packaging. Packaging of vector–insert DNA is two step processes; in the first step an extract that contains P *pac* cleavage and recognition proteins (Pacases). This extract lacks phage head, tail and accessory proteins and is produced by thermally inducing an *E. coli* strain NS3210 with a temperature sensitive P1 prophage which also carries gene 10 amber mutation. Normally, gene 10 is needed for the production of all P1 late proteins but is not necessary for the production of Pacase proteins.

In the second step of packaging; pacase enzyme recognizes and cleaves the P1 phage at the "pac" site which is then pushed into empty phage prohead, this gets condensed and then a tail is added to form complete virus. Recently, *E. coli* strain NS3690 infected with P1 phage showed higher expression of head and tail proteins and these can be easily harvested from the bacteria for using in *in vitro* packaging of recombinant P1 phages.

Materials

1. 2 to 10 ng/µl size-fractionated Sau3A or Mbo I digested genomic DNA
2. 10 to 50 ng/µl pCYPAC2 vector DNA
3. 1 Weiss U/µl T4 DNA ligase and 5X buffer
4. 0.5 M EDTA, pH 8.0
5. 10 mg/ml proteinase K
6. 100 mM PMSF solution
7. TE/PEG solution: 0.5× TE buffer, pH 8.0, containing 30% (w/v) polyethylene glycol 8000 (PEG 8000)
8. Electro competent *E .coli* DH10B

    SOC medium
9. LB plates containing 5% (w/v) sucrose and 25 µg/ml kanamycin
10. 100 × 15–mm sterile petri dishes for test transformants
11. TE buffer, pH 8.0
12. Agarose
13. 0.5X TBE (45mM Tris-borate, 1mM EDTA)
14. Submarine Electrophoresis system
15. UV - transilluminator
16. Electroporator

17. Pulsed field gel electrophoresis system

18. *E. coli* NS3208 and NS3690

19. LB agar plates and LB broth containing 25 μg/ml chloramphenicol, (30^0 to 32 ^{0}C)

20. LB broth

21. Pacase buffer [20mM Tris-HCl pH 8, 1mMEDTA, 50mM NaCl, 1mM phenylmethylsulfonyl fluoride (PMSF) prepare just before use and keep the buffer at 4 ^{0}C].

22. LB agar plates and LB medium containing 25 μg/ml chloramphenicol, and 20 μg/ml spectinomycin, warmed just above room temperature.

23. 50% (w/v) sucrose, 4 ^{0}C

24. 50 mM Tris-Cl (pH 8.0)/10% (w/v) sucrose, 4 ^{0}C

25. 10 mg/ml lysozyme solution (dissolve 10mg of lysozyme in 0.25M Tris-HCl pH 8) always prepare just before use and keep at 4 ^{0}C.

Method

A. **P1 Phage transfection:**

Materials:

i. P1 phage stock (P1 lysate)

ii. 20% glucose solution

iii. $CHCl_3$

iv. *E. coli* DH10B

v. Tris-Mg Buffer: 10mM Tris-Cl pH 7.5 and 10mM $MgSO_4$ (sterilize by Millipore filtration)

vi. 1M $CaCl_2.2H_2O$ solution (sterilize by Millipore filtration)

vii. 1M Sodium citrate solution (sterilize by Millipore filtration)

viii. P1 Salts solution: 10mM $CaCl_2.2H_2O$ and 5mM $MgSO_4$ (sterilize by Millipore filtration)

ix. Selective plates, i.e., LB agar plates with 20 μg/ml kanamycin

x. LB broth

Method

a. Inoculate *E. coli* DH10B, the recipient culture and grow overnight in 5ml LB broth at 37 ^{0}C with proper shaking (120rpm).

b. Take 1.5ml of the above culture in a sterile 1.5ml microcentrifuge tube and centrifuge at 7,000rpm for 2minutes at 4 ^{0}C.

c. Discard the supernatant and suspend the pellet in 0.75ml of P1 salts solution.

d. Take three sterile 1.5ml microcentrifuge tubes and label them as 1, 10 and 100 accordingly, add 1µl, 10µl and 100µl P1 lysate to the respective tubes. To these tubes add 100µl bacterial culture prepared in step c. As a control take another tube in which100µl of *E. coli* in P1 salts solution is added.

e. Incubate all the tubes at 37 ^{0}C for 1hour.

f. Add 1ml of LB broth and 200µl of 1M sodium citrate solution to all the tubes and incubate at 37 ^{0}C for 1hour [Note: Here sodium citrate will chelate excess calcium and facilitates absorption of phage to the *E. coli*].

g. Centrifuge all the tubes at 7,000 rpm for 2 minutes at 4 ^{0}C, discard the supernatant and suspend the pellet in 50-100µl of LB broth.

h. Spread the bacteria in selective plates in the following manner:

Incubate at 37 ^{0}C for overnight to 48hours to observe the colonies.

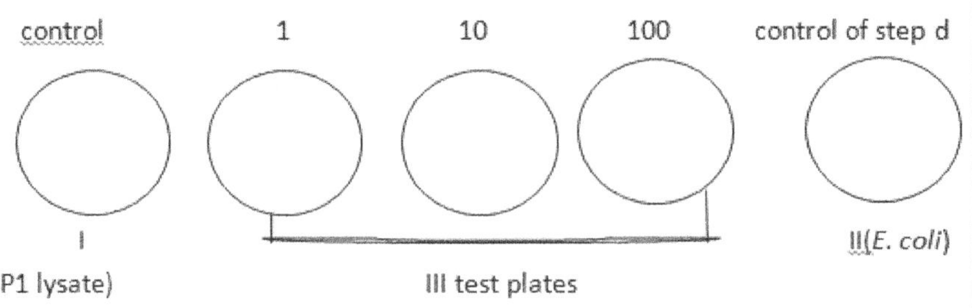

[Note: *The control plates with cells only or P1 lysate only should be free of bacterial colonies; the other plates should contain the transduced colonies. Usually the plate receiving the lowest amount of P1 lysate will have the fewest colonies*].

B. **Extraction of P1 phage DNA (pCYPAC2 vector DNA)**

Materials:

1. **TEG Buffer**

   20% Glucose

   50mM Tris-HCl, pH to 8.0

   50mM EDTA.

2. **Lysis Buffer**

   1% SDS

   0.2N NaOH

   Note: For preparing 10 ml of lysis buffer. Take 8.8 ml of sterile distilled water and to that add 0.2 ml of 10N NaOH and 1 ml of 10% SDS. (Note: Do not add SDS first as it will precipitate when NaOH is added to the solution).

3. **3M Potassium acetate pH 4.8**

   Dissolve 29.44 g potassium acetate in 25 ml of distilled water and adjust the pH to 4.8 with glacial acetic acid. Finally make up the volume to 100 ml of distilled water and sterilize by autoclaving at 15 lb / sq.in for 15 minutes.

4. **RNase. Dissolve 10 mg / ml pancreatic RNase in 50mM Tris-HCl pH 7.5 containing 0.15M NaCl. Keep the above solution in boiling water bath for 10 minutes, this is to destroy the DNase function.**

5. Isopropanol (ice cold) Chill the isopropanol by keeping it in ice before using.

6. 70% Ethanol: Mix v/v 70 ml of 100% ethanol and 30 ml of sterile distilled water.

7. 10 mg / ml Lysozyme solution. Dissolve 10 mg of lysozyme solution in 1 ml of TEG buffer and leave it on ice.

8. TE Buffer: 10mM Tris-HCl and 1mM EDTA, pH 8.0.

Method

1. Grow *E. coli* DH10B containing pCYPAC2 vector DNA in LB broth having 20µg / ml Kanamycin at 37 ^{0}C with a constant shaking at 150-200 rpm up to late log phase (OD540= 0.8 or 0.9)

2. Take 1.5 ml of the above grown culture in a microcetrifuge tube and centrifuge the cells at 7,000 rpm for 120 seconds at room temperature in a microcentrifuge.

3. Discard the supernatant and add 0.5 ml of TEG buffer and suspend the culture and centrifuge the cells at 7,000 rpm for 90 seconds at room temperature in a microcentrifuge (in case of Gram positive bacteria repeat this step thrice).

4. Discard the supernatant and suspend the pellet in 150µl of ice cold lysozyme solution and leave it in room temperature for 10 minutes.

5. To the same above mixture add 50 µl of 10mg/ml RNase and leave at 37 ^{0}C for 30 minutes.

6. Now add 300 µl of lysis buffer and mix it carefully such that clear transparent solution is formed and leave it on ice for 10 minutes.

7. Now add 250 µl of 3M potassium acetate buffer pH 4.8 slowly and leave it on ice for 30 minutes.

8. Centrifuge at 10,000 rpm for 10 minutes at 4 ^{0}C, discard the pellet and to the supernatant add 1:1 (v/v) of ice cold isopropanol mix by gentle vortexing and leave it on ice for 2-3 minutes and centrifuge at 10,000 rpm for 10 minutes at 4 ^{0}C.

9. Discard the supernatant and add 0.5 ml of 70% ethanol and centrifuge at 10,000 rpm for 10 minutes at 4 ^{0}C. Discard the supernatant and air dry the pellet.

10. Dissolve the pellet in 50µl of TE buffer. Analyze the pCYPAC2 vector DNA by running in 0.7 percent agarose gel.

C. **BamHI digestion of isolated pCYPAC2 vector DNA**

Materials:

pCYPAC2 vector DNA (1µg/10µl)

10 X BamHI buffer (New England biolabs)

BamHI (20U/µl) (New England biolabs)

Autoclaved distilled water

Calf Intestinal alkaline phosphatase (CIAP) (10U/µl) (New England biolabs)

Choloroform: isoamylalcohol solution mixed in 24:1 (v/v) (Solution B)

TE buffer: 10mM Tris-HCl and 1mM EDTA, pH 8.0.

10mM Tris-HCl

Ice cold Isopropanol

70% Ethanol

TMGD Buffer: 10mM Tris-Cl pH 8, 10mM MgCl2, 0.1% Gelatin (w/v), 10µg/ml pancreatic DNase I (this is to be added always fresh from 2mg/ml stock)

Method

1. Take 10 µl (1µg) of pCYPAC2 vector DNA in 1.5ml microcentrifuge tube and add the following: 1µl of BamHI enzyme, 10µl 10 X BamHI buffer and 79µl of autoclaved distilled water. Incubate the tube at 37 ^{0}C for 1 hour and again add 1µl of BamHI. Continue to incubate for one more hour, follow the same by adding 1µl of BamHI.

2. To the above tube add 1µl of CIAP and incubate at 37 ^{0}C for 1hour and repeat this one more time.

3. To the above tube add equal volume of solution B and mix by vortexing for 2-3 minutes at room temperature.

4. Centrifuge at 10,000 rpm for 10 minutes at 4 ^{0}C, take out the aqueous phase into a sterile 1.5ml microcentrifuge tube and add equal volume of ice cold isopropanol, mix the tubes by vortexing.

5. Perform centrifugation at 10,000 rpm for 10 minutes at 4 ^{0}C, discard the supernatant and add 250µl of 70% ethanol. Mix by vortexing and follow it up with centrifugation at 10,000 rpm for 10 minutes at 4 ^{0}C. Throw away the supernatant and air dry the pellet.

6. Suspend the pellet in 100µl of 10mM Tris-Cl pH 8.0.

D. Perform partial digestion of the isolated DNA using either Sau3A or Mbo I and analyse the generated DNA fragments on 1% agarose gel electrophoresis along with molecular size markers.

 i.   Incubate 10µg of isolated DNA

 ii. 10 X reaction buffer

 iii. Sau3A or Mbo I 5U/µl

 iv. Autoclaved distilled water so as to make up the entire volume up to 500µl.

 v.   Incubate at 37 ^{0}C

 vi. After 5 minutes of incubation please transfer 100µl from the above reaction mixture to a sterile eppendorf (1.5ml) tube containing 25µl of 0.5M EDTA, mix the solution using vortex mixer. Similarly, take out another 100µl and pipette out in another sterile autoclaved 1.5ml tube after 10 minutes. Likewise pipette out 100µl of the reaction mixture in two more tubes and finally add 25µl of 0.5M EDTA to the initial tube.

 vii. Perform 1% agarose gel electrophoresis in 1 x TAE (Tris-Acetate-EDTA pH 8.3) system along with molecular size marker.

E. Cut the agarose containing desired size of DNA fragments and electro elute the fragments.

 a.   Place the agarose containing desired fragments in dialysis bag and add 1ml of TE (10mM Tris-HCl, pH 8 and 10mM EDTA) buffer, seal both the sides with clamps and perform electrophoresis keeping the dialysis bag towards the cathode.

 b.   After 1h when the DNA fragments from gel must have moved into the buffer, stop (by observing the disappearance of orange fluorescence of ethidium bromide from the agarose gel) the electrophoresis and remove the buffer and precipitate the DNA fragments by adding equal volume of ice cold isopropanol and mix by gentle vortexing. Centrifuge at 10,000rpm for 10 minutes at 4 ^{0}C. Discard the supernatant and add 250µl of 70% ethanol and suspend the pellet by vortexing and repeat the centrifugation step. Discard the supernatant very carefully and air dry the pellet. Suspend the DNA pellet in 20µl of TE buffer.

F.　Ligation Reaction:

i.　Take 5µl BamHI digested BAC (pBACe3.6) (50ng) in a sterile 1.5ml microcentrifuge tube and to that add 2µl of 10X cohesive end ligation buffer containing 1mMATP, 2µl of insert (electroeluted DNA fragments (20ng)), 1µl of 100U/µl of T4DNA ligase and 10µl of sterile distilled water. This tube should be labelled as ligation test. Similarly, prepare ligation control in which insert DNA is absent. Incubate both the tubes at 10 ^{0}C for overnight.

ii.　Extract the DNA (test) with chloroform and isoamyl alcohol (24:1 v/v) from both ligation control and ligation test tubes by centrifuging at 10,000 rpm for 10minutes at 4 ^{0}C. Collect the aqueous phase in a sterile 1.5ml microcentrifuge tube and precipitate the DNA by adding equal volume of ice-cold isopropanol and centrifuge at 10,000 rpm for 10 minutes at 4 ^{0}C. Discard the supernatant and add 250 µl of 70% ethanol and repeat the centrifuge step. Discard the supernatant slowly and air-dry the pellet. Suspend the pellet in 10µl of TE buffer.

G.　Prepare EP cells (*E. coli* DH10B) following the protocol mentioned in section 1.1.2 (chapter 1). Add 400µl of LB media and grow the cells at 120rpm at 37 ^{0}C in a rotator orbital shaker for 1hour.

H.　Plate the cells in LB agar plate containing 20µg/ml Kanamycin and 5% sucrose. Incubate at 37 ^{0}C for overnight. Similarly, prepare for ligation control and observe the number of colonies in the ligation control which must be always by 10 to 50 folds less than that observed in ligation test plate indicating successful formation of genomic library in PAC.

I.　To identify the desired clone perform colony hybridization by using suitable probe against the target sequence.

J. Once the clone is identified, isolate the recombinant plasmid. The same can be repeated to produce P1phage lysate to store permanently the desired sequence/clone.

K. Package the ligated DNA in step F.i.

Add the following to 3 to 4 µl of the ligation mixture:

1.5 µl 100 mM Tris-Cl (pH 8.0)/250 mM NaCl/100 mM $MgCl_2$

1.5 µl 1 mM 4dNTP mix

1 µl 30 mM DTT

0.5 µl 50 mM ATP, pH 7.5

5.5 to 6.5 µl H_2O

1 µl Stage I P1 packaging extract (preparation see below).

Incubate 15 min at 30°C.

Note: *It is convenient to premix the Tris/NaCl, dNTPs, DTT, and ATP and add them to the reaction together. It is best to add the packaging extract last. The reaction should not be incubated longer than 15 min, as this may cause trace nucleases in the extract to decrease the efficiency of the process.*

L. i. Add 3 µl of 1× stage II packaging buffer and 1 µl of 50 mM ATP to the stage I reaction from step K. Pipet entire solution into a tube containing the stage II packaging extract obtained (preparation see below) (~40 to 50 µl). Mix the resulting extract, which should be viscous, with a Gilson pipettor tip and incubate 20 to 30 min at 30 ^{0}C.

ii. Terminate packaging reaction by adding 120 µl TMGD buffer and incubate 15 min at 37 ^{0}C. The reaction mixture should become significantly less viscous. Microcentrifuge reaction mixture 1 min to pellet debris. Remove phage-containing supernatant (~180 µl) to a separate tube and store at 4 ^{0}C until used. *Although the phage can be stored several weeks at 4 ^{0}C, it is best to use the preparation as soon as possible to maximize recovery of P1 clones.* Whenever the clones have to be propagated follow step A to get colonies of the desired clone.

Preparation of Stage I and Stage II P1 phage packaging extracts:

Method

I. Preparation of Stage I (Pacase) P1 phage extract

1. Streak an aliquot of *E. coli* NS3208 on a LB agar plate containing 25µg/ml chloramphenicol and incubate the plate overnight at 32 ^{0}C.

2. Transfer several colonies into tubes containing 10 ml of LB broth having 25µg/ml chloramphenicol. Continue incubation at 32 ^{0}C until cultures reach late exponential phase (OD_{540} = 1.0), ~4–5 × 10^8 cells/ml.

3. Dilute 1: 100 in LB broth and spread 20µl on each of two LB agar plates containing 25µg/ml chloramphenicol. Incubate overnight one at 32 ^{0}C and the other at 42 ^{0}C. Store the remaining cultures at 4 ^{0}C.

4. Compare the colonies obtained in LB agar plates at 32 ^{0}C and at 42 ^{0}C, the colonies in LB agar plate at 32 ^{0}C should be >100 compared with the plates growing at 42 ^{0}C. Here, *E. coli* NS3208 expresses temperature sensitive P1 prophage and those lacking this will grow at 42 ^{0}C while those bacteria expressing P1 prophage will be lysed at 42 ^{0}C.

5. Inoculate a litre of LB broth containing 25µg/ml chloramphenicol with 10 ml of the chosen culture and grow with constant shaking at 30 ^{0}C until it reaches OD_{540} = 0.5 (~4 to 5 hr). Transfer cells to a 4 ^{0}C 350ml plastic bottles and centrifuge for 10 min at 7000 rpm, 4 ^{0}C.

6. Discard supernatant and resuspend the pellet in 5 ml LB medium. Dilute the cell suspension in 1 litre of LB broth having 25µg/ml chloramphenicol and grow the culture at 42 ^{0}C with constant shaking for 15 min, lower temperature to 38°C, and continue shaking for another 165 min.

7. Chill the culture rapidly to 4 ^{0}C by shaking it several minutes in an ice-water bath. Pellet cells by centrifugation as described in step 5.

8. Resuspend the pellet in 8 ml of 4 ^{0}C pacase buffer and transfer the cell suspension to 20ml beakers. Sonicate the cell suspension with a probe having fine tip fitted to a sonicator. (Here sonication depends from sonicator to sonicator. The basic idea is to get the cells completely broken and lysate is formed).

9. Centrifuge the sonicated extracts for 30 min at 17,000 rpm at 4 ^{0}C. Distribute the supernatant as 20μl aliquots to separate 0.5-ml microcentrifuge tubes and store at −70°C. [*The extract is stable for as many as five rounds of freeze-thaw if kept at 4°C when thawed*].

II. Preparation for stage II (head-tail) extract

10. Streak an aliquot of *E. coli* NS3690 on an LB plate containing 25μg/ml chloramphenicol and 20 μg/ml spectinomycin. Incubate plate overnight at 30 ^{0}C to 32 ^{0}C and select colonies (steps 2 to 4), except use both antibiotics throughout. Grow the colonies in 10 ml of LB broth having both antibiotics. Inoculate this grown culture to 1 litre LB medium containing both antibiotics and incubate at 30 ^{0}C with constant shaking until OD_{540} = 0.3.

11. Resuspend the pellet in 5 ml LB medium and dilute in 500 ml of LB broth having 25μg/ml chloramphenicol kept at 42 ^{0}C. Incubate the culture for 20 min at 42 ^{0}C with vigorous shaking and then continue shaking at 38 ^{0}C for another 70 min.

12. In the steps outlined below, cool the cells quickly, keep all containers cold, and add sucrose to the cell suspension before centrifugation to maintain cell integrity.

13. Add 500 ml of 4 ^{0}C 50% sucrose and spin for 1 to 2 min in an ice-water bath. Pellet cells at 4 ^{0}C in 350 ml plastic bottles for 5 min at 5500 rpm.

14. Pour off all of the supernatant and blot inside bottle with a tissue to remove as much medium as possible. Resuspend the pellet in 2 ml of 4 ^{0}C 50 mM Tris-Cl (pH 8.0)/10% sucrose by gentle vortexing.

15. a) *If suspension is not viscous*: Transfer 50μl aliquots into microcentrifuge tubes containing 4μl of chilled 10 mg/ml lysozyme solution. Quick-freeze all the tubes in liquid nitrogen and store at −70°C until ready to use.

b) ***If the resuspended cell pellet is viscous*** (indicating some cell lysis): Sonicate suspension for two or three times of 10 seconds intervals at 4 ^{0}C as described in step 8, and then aliquot and freeze as explained in step 15a.

References

1. Campbell A. The future of bacteriophage biology. Nat Rev Genet. 2003;4(6):471-477.

2. Skorupski K, Pierce JC, Sauer B, Sternberg N. 1992. Bacteriophage P1 genes involved in the recognition and cleavage of the phage packaging site (pac). *J Mol Biol.* 1992; 223(4):977-989.

3. Skorupski K, Sternberg N, Sauer B. Purification and DNA-binding activity of the PacA subunit of the bacteriophage P1 pacase enzyme. *J Mol Biol.* 1994a; 243(2):258-267.

4. Skorupski K, Sauer B, Sternberg N. Faithful cleave of the P1 packaging site (pac) requires two phage proteins, PacA and PacB, and two *Escherichia coli* proteins, IHF and HU. *J Mol Biol.* 1994b; 243(2):268-282.

5. Sternberg N. A bacteriophage P1 cloning system for the isolation, amplification and recovery of DNA fragments as large as 100 kb. *Proc Natl Acad Sci U S A.* 1990; 87(1):103-7.

6. Sternberg NL. Cloning high molecular weight DNA fragments by the bacteriophage P1 system. *Trends Genet.* 1992; 8(1):11-16.

Chapter 3

CLONING IN COSMID VECTORS

COSMIDS CONSTRUCTION

Cosmids are defined as plasmids carrying "cos" site of λ phage. This was first described by John Collins in June 1978, in fact they coined the word cosmid for such cloning vehicles, this new type of vector differs from the usual plasmid vectors in having a small piece of lambda DNA, called as cohesive end site (cos site).

Cos site

Lambda DNA is linear with two single-stranded ends of twelve nucleotides in length and the ends are complementary, they are called the cohesive ends or cos sites—Right arm cos site "**GGGCGGCGACCT**" and left arm cos site "**CCCGCCGCTGGA**". Like any other replicons these cosmids also have origin of replication, selectable (antibiotic resistance) markers and gene cloning sites (multiple cloning sites) of the plasmid DNA. They lack structural and regulatory genes of λ DNA. Hence, there is no lysis and integration of cosmid DNA in the host bacteria unlike λ phage. The use of plasmid vectors in the cloning of large fragments of DNA is disadvantaged by the low transformation efficiency of large plasmids. Bacteriophage lambda can be used as a vector to clone efficiently 20-25 kilobases (kb) of foreign DNA only. Moreover large number clones are needed to be screened for assessing the clones for genome library. With the advent of whole genome sequencing of important microbial pathogens and various eukaryotic organisms including human genome, the usage of cosmids in the construction of genomic libraries and cDNA libraries increased enormously across the globe.

Materials

1. Lambda phage DNA is used as template DNA

2. Forward primer 5' **GGGCGGCGACCT**CGCGGG 3'

     -"Cos site"-

   Reverse primer 5' -TCGTTCAGCACCTGTCGT3'.

   (Specificity of the primer is defined for amplification of 136bp DNA from lambda phage DNA that contains **cos site**)

3. 10 X PCR buffer containing $MgCl_2$ (as supplied by the manufacturer)

4. *Taq* DNA polymerase (0.5-1U/µl)

5. 1mM dNTP mix

6. Autoclaved MilliQ water

7. pUC 18 (1µg/µl)

8. PfoI 10U/µl (Thermo Fisher Scientific)

9. 10X PfoI buffer (Thermo Fisher Scientific)

10. Klenow fragment 5U/µl (New England Biolabs)

11. 10 X Klenow fragment buffer (New England Biolabs)

12. T4 DNA ligase 5U/µl (Thermo Fisher Scientific)

13. 10X T4 DNA ligase for cohesive end ligation (Thermo Fisher Scientific)

14. 10X T4 DNA ligase for blunt end ligation; along with buffer add 50% PEG 4000 (Thermo Fisher Scientific)

15. *E. coli* DH5α

16. 1M $CaCl_2$. $2H_2O$ or TB

    Transformation buffer (TB): MES [2-(N-morpholino)ethanesulfonic acid] buffer contains: 10mM MES pH 6.3, 45mM $MnCl_2.4H_2O$, 10mM $CaCl_2.2H_2O$, 100mM KCl, 3mM Hexaminecobalt chloride, prepare in autoclaved distilled water. After all the salts are dissolved sterilize by Millipore filtration (filters anything between 0.45µM to 0.2µM pore size).

17. SOB Medium

     To 950 ml of deionized H_2O, add:

Bacto tryptone	20 g
Bacto yeast extract	5 g
NaCl	0.5 g

     Shake until the solutes have dissolved. Add 10 ml of a 250mM solution of KCl. (This solution is made by dissolving 1.86 g of KCl in 100 ml of deionized water). Adjust the pH to 7.0 with 5N NaOH (0.2 ml). Adjust the volume of the solution to 1 litre with deionized H_2O. Sterilize by autoclaving for 20 minutes at 15 lb / sq in on liquid cycle.

18. LB-Agar.

19. 2M Glucose (sterilize by Millipore filtration)

20. 2M $MgCl_2$.$6H_2O$ (2M $MgCl_2$: This solution is made by dissolving 19 g of $MgCl_2$ in 90 ml of deionized H_2O. Adjust the volume of the solution to 100ml with deionized H_2O and sterilize by Millipore filtration).

21. 100mg/ml Ampicillin dissolved in autoclaved distilled water

22. 1M IPTG solubilized in autoclaved distilled water.

23. 50mg/ml X-Gal dissolved in N N' dimethyl formamide.

24. **SOC Medium**: SOC Medium is identical to SOB medium, except that it contains 20mM glucose and 20mM of $MgCl_2$. Inoculate the media with 0.5ml of mid log phase bacterial culture and allow it to grow until OD540nm is between 0.5 – 0.6.

25. **Luria - Bertani (LB) Medium**

Bacto tryptone	10g
Bacto yeast extract	5g
NaCl	5-10g
dH_2O	1 litre

     Adjust the pH to 7.0 with 10 N NaOH (0.4ml/litre).

26. **LB agar plates:** 15 g of Bacto agar/litre LB media.

27. 10% Glycerol: 10ml of autoclaved glycerol dissolved in 70ml of autoclaved distilled water and after dissolution the volume is made up to 100ml with autoclaved distilled water.

28. Solution B: $CHCl_3$: isoamylalcohol (24:1 v/v)

29. Calf intestine alkaline phosphatase (CIAP) 10U/µl

Method

1. The reaction mixture is contained in a final volume of 50µl which consisted of 500ng of isolated DNA, 50pM of each primers, 100µM of dNTPS mix, 1X reaction buffer that contained $MgCl_2$ supplied by the manufacture, 1U of hot start *Taq* DNA Polymerase.

2. The amplification parameters were calibrated in a Mastercycler gradient Thermocycler gradient (Eppendorf) which included initial denaturation for 10 min at 94 ^{0}C, followed by forty cycles of 94 ^{0}C for one minute of denaturation, 60 sec of annealing at 55 ^{0}C and amplification at 72 ^{0}C for 30 sec, the cycles ended with final reaction at 72 ^{0}C for 5 minutes.

3. The PCR product is analyzed in 1% agarose gel electrophoresis. Here include the template DNA and load in one well.

4. Record the gel and cut the desired range of DNA fragments gel and electroelute the fragments in 10mMTris-HCl, 10mM EDTA pH 8.0 buffer present in the dialysis bag. Perform electroelution for 45 minutes to 60 minutes by keeping the bag at cathode.

5. Take out the buffer containing DNA fragments and add equal volume of ice cold isopropanol and mix the solutions briefly by vortexing; following this centrifuge the mixture at 10, 000 rpm for 10 minutes at 4 ^{0}C.

6. Discard the supernatant slowly and add 500µl of 70% ethanol. Vortex briefly and perform the centrifugation as mentioned in step 5.

7. Discard the supernatant very slowly and air dry the pellet

8. Once all the alcohol is evaporated suspend the pellet in 50µl of TE buffer and use it for cloning in the plasmid for the generation of cosmid.

9. Take 10µl of PCR products explained in step 8 in a sterile 1.5ml microcentrifuge tube and add 5µl of 10 X Klenow fragment buffer, 5µl

of 1mM of dNTP mix, 1μl of 5U/μl Klenow fragment and make up the volume to 50 μl with autoclaved milliq water. Incubate at 37 ^{0}C for 30 minutes.

10. Add 50 μl of solution B to the above tube and vortex for 1-2minutes at room temperature and centrifuge at 10,000rpm for 10 minutes at 4 ^{0}C.

11. Take out the aqueous phase in a sterile 1.5ml microcentrifuge tube and add equal volume of ice-cold isopropanol mix by gentle vortexing and perform centrifugation as explained in step 10.

12. Discard the supernatant and add 250μl of 70% ethanol and repeat centrifugation step 10. Discard the supernatant and air dry the pellet; following this dissolve the pellet in 20μl of 10mM Tris-HCl pH 8 and use it for ligation into pUC 18 vector linearized with PfoI enzyme.

13. PfoI enzyme treatment of pUC18: 1μl of pUC 18 was mixed with 5μl of 10X PfoI enzyme buffer, 1μl of PfoI enzyme, 43μl of sterile milliq water and incubate at 37 ^{0}C for 1h. This step is repeated for two more times. This is followed by the addition of 1μl of CIAP and incubation at 37 ^{0}C for 1h and this step is repeated for one more time.

14. To the above tube add equal volume of solution B and vortex briefly and centrifuge as mentioned in step 10. After this follow steps 11 and 12. Here dissolve the pellet in 50μl of 10mM Tris-HCl pH 8. Use this in the ligation of PCR product blunt ended as mentioned in step 9.

15. Take sterile 1.5ml microcentrifuge tube and add 10μl of pUC18 linearized with PfoI and 2.5μl of insert (136 bp PCR product containing cos site of λ phage DNA), 2μl 10X T4DNA ligase blunt end buffer, 2μl of 50% PEG 4000, 1μl of T4 DNA ligase (100U) and 2.5μl sterile milliq water. Similarly, keep ligation control without adding the insert and increase sterile milliq water to 5μl. Incubate at 10 ^{0}C for overnight.

16. Transformation:

    i.   Streak *E. coli* DH5α cells on LB-agar plate and incubate at 37 ^{0}C for overnight. Pick a single colony from the plate and grow up to OD_{540} = 0.6 (10 hrs growth).

    ii. Inoculate 50 ml of this culture to 100 ml of SOB media containing 20mM of glucose and 20mM of $MgCl_2.6H_2O$ and grow at 37 ^{0}C for 5 hrs with constant shaking at 250 rpm (such that $O.D_{540}$ = 0.5).

iii.   Leave the grown *E. coli* DH5α culture on ice for 30 minutes. Pellet the cells by centrifuging at 7,000 rpm for 10 minutes at 4^0C.

iv   Dilute the stock 1M calcium chloride to 0.1M $CaCl_2$. $2H_2O$ and chill it by keeping it in ice and suspend the pellet in 10 ml of chilled 0.1M $CaCl_2$. $2H_2O$, or in TB. Leave it in ice for 30 minutes. *[Note: **This is 1/10 volume of the bacterial culture volume]**.*

v.   Centrifuge the cells at 7,000 rpm for 10 minutes at 4^0C. Discard the supernatant very carefully as the pellet would be loose and suspend the pellet in 1 ml of ice cold 0.1M $CaCl_2$. $2H_2O$ or TB. At this stage these cells are called as competent cells and they can be stored in 25% v/v glycerol (750μl competent cells and 250μl 100% sterile glycerol) and kept at -20 ^{0}C.

vi.   Take 200 ml of the competent cells in a sterile 1.5ml microcentrifuge tube and add 10 ml of the plasmid (containing 1 to 10ng of the DNA) and leave it on ice for 30 minutes. (In case of cloning of a DNA fragment or PCR product or cDNA or genomic library preparation prepare 5 tubes containing 200 ml of the competent cells and to that add one DNA control mentioned in Table 1 such that you have four controls and one tube contains the ligation mixture).

vii. Keep the above tubes at 42 ^{0}C for 90 seconds and then immediately leave them on ice for 2 to 3 minutes.

viii. Add 800 ml of LB media and grow at 37 ^{0}C for 1hr with constant shaking at 250 rpm.

ix.   Plate LB-Agar plates with 100mg/ml ampicillin, 1mM IPTG and 20μg/ml X-gal, and incubate at 37 ^{0}C for overnight, following day screen the plates to establish the successful transformation taking into consideration of controls (Table 1 and eq.1). Once the recombinants are ascertained prepare for screening the desired clone.

Table 1 Different controls used to screen recombinant clones

Control	Selection pressure	Growth
Insert control	Ampicillin+ve, IPTG and X-gal	-ve
Vector control	Ampicillin +ve, IPTG and X-gal	+ve
Ligation control	Ampicillin +ve, IPTG and X-gal	+ve
Positive cell control	Ampicillin +ve, IPTG and X-gal	-ve
Only competent E. coli DH5α bacteria were plated		
Negative cell control	No selection pressure	+ve
Only competent E. coli DH5α bacteria were plated		

Based upon these controls whether the cloning is successful or not can be ascertained in the test plate. The number of colonies in the test plate must be at least five to six folds higher than that present in the ligation control to accentuate successful formation of recombinants. We calculate the transformation efficiency as follows:

$$\text{Transformation efficiency} = \frac{\text{No. of transformants (colonies) x Final volume at recovery (ml)}}{\mu\text{g of plasmid DNA x volume plated (ml)}} \quad \text{eq.1}$$

17. The positive clones are grown in the presence of 100μg/ml ampicillin for overnight and the cosmid DNA is isolated by the alkaline lysis method mentioned in section 1.2.3 of Chapter 1 of this book.

18. Confirm the presence of 136bp fragment by taking the isolated cosmid as template and perform PCR using the primers and condition mentioned in section 3.1 of this chapter. Resolve the PCR products in 2% agarose gel electrophoresis performed in 1x TAE (Tris-acetate EDTA) system.

PREPARATION OF GENOMIC LIBRARY IN COSMID VECTORS

Partial Restriction Digestion of Chromosomal DNA:

A. **Sau3A digestion of chromosomal DNA**

Materials:

10X reaction buffer as supplied by the manufacturer

20 Unit/µl of Sau3A or as defined by the manufacturer

0.5M EDTA, pH 8

50X TAE buffer pH 8.3

Low EEO (electro endosmosis) agarose

Ice cold isopropanol

70% ethanol: 30 ml autoclaved distilled water and 70 ml 100% ethanol.

Human Chromosomal DNA

Method

1. Take 5µg (25µl) of DNA isolated from the organism in a sterile 1.5 ml microcentrifuge tube.

2. Add 50 µl of 10X reaction buffer

3. Add 425 µl of autoclaved distilled water

4. 2µl of Sau3A (20 Unit/µl)

5. Take 5 sterile 1.5ml microcentrifuge tubes and label them as 5', 10' 20' and 40' and add 50µl of 0.5M EDTA.

6. Incubate the above mixture (steps 1 to 4) at 37 ^{o}C. Exactly after 5' pipette 100µl of the reaction mixture and place it in 5' labelled 1.5 ml microcentrifuge tube and vortex briefly and leave at room temperature.

7. Similarly, pipette out 100µl after 10', 20' and 40' and transfer it to the respective tubes containing EDTA.

8. After 60 minutes to the remaining 100µl add 50µl of 0.5 MEDTA vortex and leave at room temperature.

9. Perform 1% agarose gel electrophoresis along with Supermix DNA molecular size ladder and undigested DNA in 1X TAE buffer under submerged conditions.

10. Record the gel and cut the desired range of DNA fragments gel and electroelute the fragments in 10mMTris-HCl, 10mM EDTA pH 8.0 buffer present in the dialysis bag. Perform electroelution for 45 minutes to 60 minutes by keeping the bag at cathode.

11. Take out the buffer containing DNA fragments and add equal volume of ice cold isopropanol and mix the solutions briefly by vortexing, following this centrifuge the mixture at 10, 000 rpm for 10 minutes at 4 ^{0}C.

12. Discard the supernatant slowly and add 500µl of 70% ethanol. Vortex briefly and perform the centrifugation as mentioned in step 11.

13. Discard the supernatant very slowly and air dry the pellet

14. Once all the alcohol is evaporated suspend the pellet in 100µl of TE buffer and use it for the construction of genomic library.

B. **BamHI digestion of Cosmid**

Materials

1. 10X reaction buffer as supplied by the manufacturer

2. 20 Unit/µl of Bam HI or as defined by the manufacturer

3. 20 Unit/µl Sma I or as defined by the manufacturer

4. 20 Unit/µl EcoRI or as defined by the manufacturer

5. 0.5M EDTA, pH 8

6. 50X TAE buffer pH 8.3

7. Low EEO agarose

8. Ice cold isopropanol

9. 70% ethanol: 30 ml autoclaved distilled water and 70 ml 100% ethanol

10. Cosmid vector DNA 1µg/µl

11.   Calf intestine alkaline phophatase (CIAP) (10Units/µl)

12.   Solution B: Chloroform:isoamylalcohol mixed in 24:1 (v/v).

Method

1. Take 1µg (10µl) of cosmid vector DNA isolated in above section in a sterile 1.5ml microcentrifuge tube.

2. Add 5 µl of 10X reaction buffer

3. Add 84 µl of autoclaved distilled water

4. 1µl of Bam HI (20 Unit/µl) and incubate at 37 ^{0}C for 1 hour and repeat this step two more times.

5. Add 1µl of CIAP (10 Unit/µl) to the above solution and incubate at 37 ^{0}C for 1 hour and repeat this step two more times.

6. Add equal volume of solution B and vortex for 2 minutes at room temperature and perform centrifugation at 10, 000 rpm for 10 minutes at 4 ^{0}C.

7. Take out the clear aqueous phase in a sterile 1.5ml microcentrifuge tube and add equal volume of ice cold isopropanol and vortex briefly. Perform the centrifugation at 10, 000 rpm for 10 minutes at 4 ^{0}C.

8. Discard the supernatant very slowly and add 250 µl of 70% ethanol and vortex for 2 minutes and perform centrifugation at 10, 000 rpm for 10 minutes at 4 ^{0}C.

9. Discard the supernatant very slowly and air dry the pellet.

10. Dissolve the pellet in 100µl of TE buffer (10mMTri-HCl and 1mM EDTA pH 8).

C. **Construction of genomic library in cosmid vector:**

Materials

1.   20% Maltose: Dissolve 20g of maltose in 100ml distilled water; sterilize by 0.22 micron Millipore filtration.

2.   100mg/ml Ampicillin dissolved in autoclaved distilled water.

3.   1M IPTG solubilized in autoclaved distilled water.

4. 20mg/ml X-Gal (20mg of X-Gal = 5'Bromo 4'Chloro 3' indolyl β-D galactoside dissolved in 1ml of N' dimethylformamide).

5. **SOC Medium**: SOC medium is identical to SOB medium, except that it contains 20mM glucose and 20mM of $MgCl_2$. Inoculate the media with 0.5ml of mid log phase bacterial culture and allow it to grow until OD540nm is between 0.5 – 0.6.

6. **Luria - Bertani (LB) Medium**

Bacto tryptone	10g
Bacto yeast extract	5g
NaCl	5-10g
dH_2O	1 litre

   Adjust the pH to 7.0 with 10 N NaOH (0.4ml/litre).

7. **LB agar plates:** 15 g of Bacto agar/litre LB media.

8. **10% Glycerol:** 10ml of autoclaved glycerol dissolved in 70ml of autoclaved distilled water and after dissolution the volume is made up to 100ml with autoclaved distilled water.

9. λ control DNA.

10. SM buffer: 0.1 M NaCl, 0.008 M $MgSO_4$, 0.01% gelatin. Sterilize by autoclaving

Method

1. Genomic DNA fragments are generated by partial digestion of chromosomal DNA with Sau3A to produce fragments 37–48 kb in length as generated in section 3.2.

2. The genomic insert DNA is ligated to the linearized vector DNA, producing concatemeric hybrid structures following step 15 of section 3.1 of this chapter.

3. Ligated DNA is mixed with λ bacteriophage packaging extract. Molecules with cos sites, approx 37–48 kb apart, will be packaged into phage heads. Take 1:10 ratio of ligate mixture and λ bacteriophage packaging extract (v/v) in a sterile 0.5ml microcentrifuge tube.

4. Add 200 µl of *E. coli* DH5α (competent cells prepared as mentioned in section 1.2.3) and 200 µl plain *E. coli* DH5α culture (grown in LB broth containing 2% Maltose) in sterile 5ml glass tube and add the solution at step 3. Continue to add 3 µl of 1M IPTG and 30 µl of X-Gal and add 2.66 ml of molten top agar maintained at 50 ^{0}C and quickly pour on a LB agar plate and allow the agar to solidify and keep the plate at 37^0 C for 8-10 hours and observe light blue coloured colonies and white colonies in the plate.

5. *E. coli* DH5α strains are infected with the packaged phage and the recombinant cosmid is propagated as a large plasmid in the host cells and screened using ampicillin, IPTG and X-gal in the LB agar plate. The white colonies are recombinants and blue colonies are non-recombinants.

6. The cosmid library can then be screened appropriately for clones of interest using specific probes, following colony hybridization methodology as described in chapter 1.

    Note: If the cosmid does not contain lac Z gene and its regulatory elements then follow the procedure described below.

7. The day before the infection procedure, streak the appropriate E. coli host strain from a glycerol freezer stock onto appropriate media and incubate at 37 ^{0}C overnight.

8. Inoculate a single bacterial colony into 15 mL of LB media, supplemented with $MgSO_4$ and maltose, in a 50-mL conical tube.

9. Incubate the culture, with shaking at approx. 200 rpm, until late log-phase growth (optical density [OD600] = 1.0), either at 37 ^{0}C for 4–6 h, or overnight at 30 ^{0}C.

10. Centrifuge the culture at 7000 rpm for 10min at 4 ^{0}C to pellet the bacterial cells.

11. Gently resuspend the cells in 7.5 mL of sterile 10 mM MgSO4. Dilute the bacterial cells to an OD600 of approx 0.5 with sterile 10 mM $MgSO_4$.

12. Prepare a 1 : 10 dilution of the cosmid packaging reactions in SM buffer. Mix 25 µL of each dilution with 25 µL of prepared bacterial host strain in a 1.5-mL microcentrifuge tube. Incubate the tubes at room temperature for 30 min.

13. Add 200 µL of LB broth and incubate at 37 ^{0}C for 1 h. Gently shake the tube every 15 min to mix the cells.

14. Centrifuge the samples for 30–60 s in a microcentrifuge to pellet the cells. Gently resuspend the cell pellet in 50 µL of fresh LB broth.

15. Plate the cells on LB agar plates containing the appropriate antibiotic selection. Incubate the plates at 37 ^{0}C overnight.

PARTIAL cDNA LIBRARY CONSTRUCTION

Preparation of cDNA

Follow the protocol mentioned in chapter 2 and clone in cosmid here if the size of the recombinant is less than 37Kb follow transformation procedure as mentioned in chapter 1 and if the size of the recombinant molecule is between 37-52Kb follow the above mentioned protocol. The obtained colonies can be screened following colony hybridization technique described in chapter 1.

References

1. Collins J, Braining HJ. Plasmids useable as gene-cloning vectors in an *in vitro* packaging by coliphageλ: "Cosmids". *Gene*.1978; 4: 85-107.

2. Collins J, Hohn B. Cosmids: A type of plasmid gene-cloning vector that is packageable *in vitro* in bacteriophage λ heads. *Proc. Natl. Acad. Sci. USA*. 1978; 75(9):4242-4246.

3. Sambrook J, Russell DW, *Molecular cloning*. A laboratory manual, Cold Spring Harbor Laboratory Press, Cold Spring Harbor, New York 2001.

4. Miki T, Matsui T, Heidaran MA, Aaronson SA. An efficient directional cloning system to construct cDNA libraries containing full-length inserts at high frequency. *Gene*. 1989; 83: 137-146.

5. Feiss M, Winder W. Bacteriophage λ DNA packaging: Scanning for the terminal cohesive end site during packaging (terminase). *Proc. NatL Acad. Sci. USA*. 1982; 79: 3498-3502.

6. Nichols BP, Donelson JE. 178-Nucleotide Sequence Surrounding the cos Site of Bacteriophage Lambda DNA. *J.Virol.* 1978; 26(2): 429-434.

7. White MJA, Nichols WA. Cosmid Packaging and Infection of *E. coli*. In *Methods in Molecular Biology*, Vol. 235: *E. coli* Plasmid Vectors. Edited by: N. Casali and A. Preston © Humana Press Inc., Totowa, NJ.

8. Brady SF. Construction of soil environmental DNA cosmid libraries and screening for clones that produce biologically active small molecules. *Nature Protocols.* 2007; 2(5):1297-1305.

Chapter 4

TECHNIQUES TO UNDERSTAND THE FUNCTIONALITY OF A GENE

PHAGE DISPLAY METHOD

Background

Phage display is a method employed to study the protein–protein, protein–peptide, and protein–DNA interactions using bacteriophages (M13 or λ) to relate proteins with the exact codons encoding the gene.

In this procedure, a gene encoding a protein of interest is cloned into a phage genome replacing its coat protein gene thus, making the phage to "display" or "express" the protein of interest on the phage coat exactly resembling the gene present in the recombinant phage genome, thereby drawing exact relationship between genotype and phenotype. In this fashion the displaying phages can be screened with antibodies raised against native protein or its peptide. This technique makes to construct large protein or peptide libraries which can be screened and amplified, this *in vitro* selection is similar to natural selection. In this method, M13mp vectors are best suited for successfully displaying the cloned gene in place of coat protein gene.

Phage display is used for the high-throughput screening of protein interactions and M13 filamentous phage are the most used vectors that display the DNA encoding the protein or peptide of interest is ligated into the pIII or pVIII gene, which forms tail and major coat proteins respectively. The phage gene and insert DNA hybrid is then inserted (a process known as "transduction") into *Escherichia coli* bacteria such as TG1,

SS320, ER2738, or XL1 that expresses correctly as fusion protein of pIII or PVIII proteins. Most of the display technique is performed exploiting the pIII gene and in this process the phage becomes incapable of getting packaged into phage. To overcome this; such recombinant phages are expressed in *E. coli* containing F plasmid and helper M13 phage which plays vital role in packaging the recombinant phages also.

A phagemid or phasmid is a plasmid that contains an f1 origin of replication from an f1 or from M13 phage making the molecule to replicate as a plasmid, and also be packaged as single stranded DNA in viral particles. Phagemids are used to clone DNA fragments and transformed into host *E. coli* using calcium chloride mediated transformation or through electroporation. The *E. coli* host contains a helper M13 phage and F plasmid allowing the phagemid to be packaged as single stranded phage like M13 phage. In this process the recombinant pIII protein in the recombinant phagemid will be expressed in the tail and can be identified with suitable antibody probe (Figure 1).

This phage display can be used to generate several protein or peptide or antibody libraries; and has become a powerful method for studying both the immune response as well to rapidly select and evolve human antibodies for therapy.

PHAGE DISPLAY

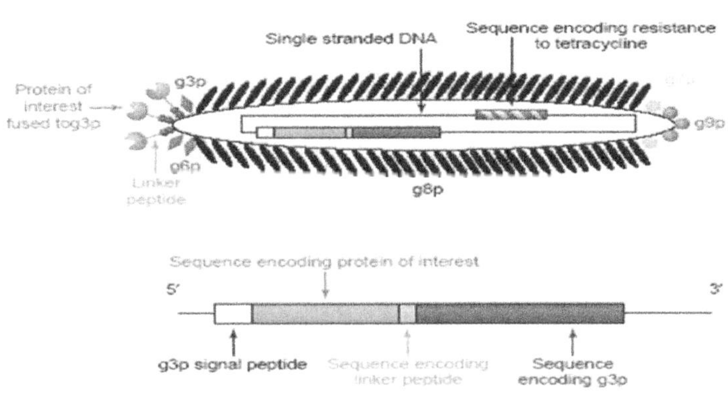

Figure 1 Schematic representation of phage display technique

Materials

1. **Transfection of M13mp18 phage into *E. coli* JM109/TG1. Follow as described in chapter II.**

2. Isolation of double stranded and single stranded M13mp18 phage DNA, pursue as explained in chapter II.

3. BsmI digestion of M13mp18 and end filling with Klenow fragment

Materials

BsmI (New England Biolabs) 10U/µl

10 X Bsm I digestion buffer

Autoclaved distilled water

M13mp18 (1µg)

Solution B: $CHCl_3$:isoamylalcohol (24:1 v/v)

Ice cold isopropanol

70% ethanol

T Buffer: 10mMTris-HCl pH 8

1mM dNTP mix

Klenow fragment (1U/µl)

10X Klenow buffer

Method

i. Take sterile 1.5ml microcentrifuge tube and take 10µl (1µg) of M13mp18 DNA and add the following: 5µl of 10X Bsm I digestion buffer, 34µl of autoclaved distilled water, 1µl of Bsm I and incubate at 37 ^{0}C for 1 hour. Add 1µl of Bsm I and incubate 1 hour more to ensure maximum M13 molecules are linearized. Sometimes it is advisable to perform addition of restriction enzyme three times to achieve maximum results.

ii. To the above tube add equal volume of solution B and vortex for 2-3 minutes in a vortex mixer and centrifuge at 10,000rpm for 10minutes at 4 ^{0}C.

iii.   Decant aqueous phase into a sterile 1.5ml microcentrifuge tube and add equal volume of ice cold isopropanol, mix for 1minute in a vortex mixer and centrifuge at 10,000rpm for 10minutes at 4 ^{0}C.

iv.   Discard the supernatant and add 250µl of 70% ethanol and centrifuge at 10,000rpm for 10minutes at 4 ^{0}C. Completely discard the supernatant and air dry the pellet.

v.   Dissolve the pellet in 10µl of T buffer, add 10µl of 1mMdNTP mix, 5µl of 10X Klenow fragment buffer and 1µl of Klenow fragment and incubate at 37 ^{0}C for 1h. Extract the DNA following steps II to IV. This is ready for ligation to PDF gene of *Staphylococcus aureus.*

4.   Preparation of PDF and cloning into Bsm I linearized M13mp18

i.   *Amplification of Peptidyl deformylase (PDF) from Staphylococcus aureus*

Materials

Chromosomal DNA of *Staphylococcus aureus* (1µg)

Forward Primer: 5' TATGTTAACAATGAAAGAC 3' and Reverse Primer: 5' TACAGCATCTGTATGTGG 3'for the Polymerase chain reaction (PCR) to amplify PDF gene

PCR conditions: 94°C for 10 min, 94°C for 45 s, 45°C for 45 s, 72°C for 2 min, extension 72°C, 5 min

10 X PCR buffer containing $MgCl_2$ (as supplied by the manufacturer)

Taq DNA polymerase (0.5-1U/µl)

T Buffer: 10mMTris-HCl pH 8

Method

1.   The reaction mixture is contained in a final volume of 50µl which consisted of 500ng of isolated DNA, 50pM of each primers, 100µM of dNTPS mix, 1X reaction buffer that contained $MgCl_2$ supplied by the manufacturer, 1U of hot start *Taq* DNA Polymerase.

2. The amplification parameters were calibrated in a Mastercycler gradient Thermocycler gradient (Eppendorf) which included an initial denaturation for 10 min at 94 °C, followed by forty cycles of 94 °C for 45 sec of denaturation, 45 sec of annealing at 45 °C and amplification at 72 °C for 20 sec, the cycles ended with final reaction at 72 °C for five minutes.

3. The PCR product is analyzed in 1% agarose gel electrophoresis. Here include the template DNA and the test sample. Cut the agarose carrying the desired size of the DNA fragment (PDF PCR product) and electroelute the DNA as described in chapter 1. The pellet is suspended in 10μl of T buffer. Follow the blunt end generation given in step v.

ii. Ligation Reaction with T4 DNA ligase

Materials

10X Blunt end ligation buffer

T4 DNA ligase (100U/μl)

Sterile distilled water

Method

a. Take 5μl of M13mp18 digested with Bsm I and ends filled with Klenow fragment obtained in step v in a 1.5ml sterile microcentrifuge tube, add 1μl of PDF gene obtained in step 3, 2μl of 10X blunt end ligation buffer, 11μl of sterile distilled water and 1μl of T4DNA ligase. The generated recombinant molecule is referred as phagemid (phPDF-M13gIII). Similarly prepare another tube which does not contain the insert PDF gene which serves as ligation control. Incubate both the tubes at 15 °C overnight.

5. **Transformation of phagemid (phPDF-M13gIII) into *E. coli* JM109/TG1, pursue as mentioned in chapter II. Screen the plates without adding X-gal and IPTG. For ascertaining the successful cloning, check the number of plaques formed in ligation control and ligation mixture. Pick up the recombinant clones from ligation mixture plate and propagate in the host *E. coli* and isolate the double stranded phagemid- phPDF-M13gIII as mentioned in chapter II.**

6.   Phage Diplay of phPDF-M13gIII.

Materials

E. coli JM109/TG1 containing M13mp18 which serves as helper phage, transformation experiment follow as mentioned in **chapter 1**.

Antibodies against *S. aureus* PDF enzyme raised in rabbit

Antirabbit IgG coupled to Horse radish peroxidise

10 X Tris HCl stock: 1MTris-HCl pH 8 and 3M NaCl

Wash Buffer: Dilute 1:10 10 X Tris-HCl stock, 0.3 % (v/v) Tween - 20 and 0.05% (v/v) Triton X-100.

Blocking Buffer: Add 5% blotto (5 g non-fat dry milk to 100 ml wash buffer). Alternatively, use 3% BSA in wash buffer.

Method

1. The method till picking up of the plaques on to nitrocellulose membrane is same as explained in above section phage display.

2. Carefully lift the membrane such that plaques stick to the membrane.

3. Incubate the NCM in blocking buffer for 1h at 37 ^{0}C and then perform washing (1 x 5mintues with gentle agitation) 6 times with 30 ml of wash buffer.

4. Incubate the NCM in primary antibody at 37 ^{0}C for 1 hr. Wash the NCM 6 times with wash buffer.

5. Add secondary antibody (anti rabbit) conjugated to horse radish peroxidase. Incubate at 37 ^{0}C for 1 hr followed by 6 times washing with wash buffer.

6. Pour off the last wash buffer and add 20 ml of developing reagent. (6 mg of 3, 3' diaminobenzidine tetrahydrochloride in 10 ml of 0.01 M Tris-HCl pH 8.0 and 30 μl of 30% H_2O_2) and observe the spots specific to the expressed antigen in the recombinant clones.

7. Place the LB plate containing the positive plaques and pick up positive clone with the help of a tooth pick and put it in a LB tube where *E. coli* host is growing at mid log phase (OD_{540} = 0.5). Grow for another 12 hours at 37 ^{0}C.

8. Centrifuge the tube at 7,000rpm for 10 minutes at 4 ^{0}C, the supernatant contains the pure recombinant clone in single stranded form which can be preserved in 30% glycerol and stored indefinitely at -20 ^{0}C.

9. The pellet can be used to isolate double stranded phagemid and M13mp18 DNA by following the procedure explained in chapter II.

References

1. Smith GP. Filamentous fusion phage: novel expression vectors that display cloned antigens on the virion surface". *Science*. 1985; 228 (4705): 1315–1317.

2. Smith GP, Petrenko VA (1997). Phage Display. Chem. Rev. 1997; 97 (2): 391–410.

3. Kehoe JW, Kay BK. Filamentous phage display in the new millennium. *Chem. Rev.* 2005; 105 (11): 4056–4072.

4. Lunder M, Bratkovic T, Doljak B, Kreft S, Urleb U, Strukelj B, Plazar N. Comparison of bacterial and phage display peptide libraries in search of target-binding motif. *Appl Biochem Biotechnol.* 2005; 127(2):125-31.

5. Chasteen L, Ayriss J, Pavlik P, Bradbury AR. Eliminating helper phage from phage display. *Nucleic Acids Res.* 2006;34(21):e145.

6. Sambrook J, Russell DW, *Molecular cloning.* A laboratory manual, Cold Spring Harbor Laboratory Press, Cold Spring Harbor, New York 2001.

7. Hufton SE, Moerkerk PT, Meulemans EV, de Bruïne A, Arends JW, Hoogenboom HR. Phage display of cDNA repertoires: the pVI display system and its applications for the selection of immunogenic ligands. *J Immunol Methods.* 1999; 231(1-2):39-51.

8. Hammers CM, Stanley JR. Antibody Phage Display: Technique and Applications. *J Invest Dermatol.* 2014;134(2):1-5.

9. Pini A, Viti F, Santucci A, Carnemolla B, Zardi L, Neri P, Neri D. Design and use of a phage display library. Human antibodies with sub-nanomolar affinity against a marker of angiogenesis eluted from a two-dimensional gel. *J Biol Chem.* 1998;273(34):21769-21776.

10. Arap MA. Phage display technology - Applications and innovations. *Genetics and Molecular Biology.* 2005; 28(1): 1-9.

11. Bazan J, Całkosiński I, Gamian A. Phage display--a powerful technique for immunotherapy: 1. Introduction and potential of therapeutic applications. *Hum Vaccin Immunother.* 2012; 8(12):1817-1828.

SITE DIRECTED MUTAGENESIS (SDM):

In site-directed mutagenesis, a specific or intentional change in the DNA sequence of a gene is to understand the functionality of the gene product both *in vitro* and *in vivo*. Sometimes it is known as site-specific mutagenesis or oligonucleotide-directed mutagenesis. Site-directed mutagenesis is one of the most important techniques in laboratory for introducing a mutation into a DNA sequence. However, with decreasing costs of oligonucleotide synthesis, artificial gene synthesis is now occasionally used as an alternative to site-directed mutagenesis.

The gene is first cloned in the M13mp18 vector and tansfected into *E. coli* TG1 or JM109, and the positive plaque is picked with the help of a sterile tooth pick and put into a tube of LB broth in which *E. coli* TG1 or JM109 bacteria is growing (OD_{540}=0.5). Allow the tube to grow for at least 6-8 hours at 37 ^{0}C. Centrifuge the tube at 10,000 rpm for 15 minutes at 4 ^{0}C and the supernatant contains single stranded M13mp18 which also has the desired gene. From the phage, the single stranded (ss) DNA is extracted and used in the incorporation of mutations in the gene.

The main criterion in the SDM is one should know the complete gene sequence and amino acid sequence of the gene product. Recently, the gene sequence can be directly annotated to amino acid sequence and thereby through cloning in correct frame in any expression vector one can obtain the gene product which can be easily characterized. As a first step the active site region of the gene product such as substrate binding site, or ATP binding site or cofactor factor binding site, etc., all of them that influence the functions of the gene product are first deciphered. An oligonucleotide primer is designed having the mutation correctly incorporated in the second base of the active amino acid(s) which ensures incorporation of new amino acid in the gene product. This primer is added to the mixture that contains ssDNA of M13mp18 having the desired gene, dNTP's and buffer containing Mg^{2+} finally, T7 polymerase enzyme is added and incubated at 37 ^{0}C for 1 hour. This is again transfected into *E. coli* TG1 or JM109, the desired clone is picked and double stranded M13mp18 having the desired gene is isolated from the *E. coli* TG1 or JM109. The mutated gene is amplified using the desired gene specific primers and the PCR product is cloned in an expression vector in correct frame so as to get the desired gene product. The product is then characterized and compared with native gene product. This makes us to understand the involvement of those amino acids in the active functioning of the gene in the organism (Figure 2).

The single stranded M13mp18 DNA having the desired gene can be directly used in the Sanger's method of DNA sequencing or in the dye terminating DNA sequencing. Here, universal primer of the M13mp18 can be used to sequence the gene directly.

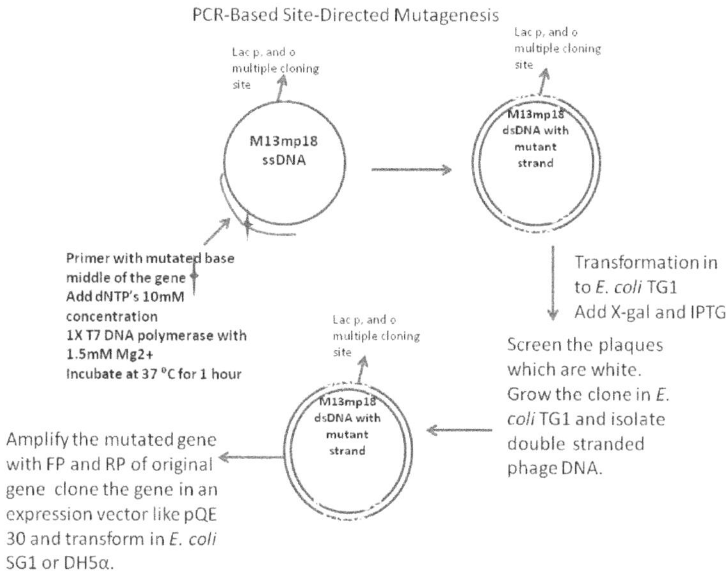

Figure 2 Schematic representation of site-directed mutagenesis using M13mp18 vector. Instead of M13mp18 vector, pUC18 vectors can also be used to carry out site-directed mutagenesis.

Materials

1. M13mp18 or pUC18
2. Sma I (10U/μl)
3. *Staphylococcus aureus* chromosomal DNA
4. Forward and Reverse Primers of glkA gene of *Staphylococcus aureus*
5. Mutated primer (where second base of a triplet code coding for an amino acid is replaced such that in place of original amino acid a new amino acid is incorporated which has profound effect on the protein functioning)
6. T4 DNA polymerase (Sequenase)
7. 10 X T4 DNA polymerase buffer with Mg^{2+} ions
8. 10mM dNTPs (dATP, dGTP, dTTP and dCTP)
9. Incubator 37 ^{0}C.
10. *E. coli* TG1
11. X- gal (5'Bromo 3' chloro 3' indolyl β-Galactoside)
12. IPTG (Isopropyl βD-galactoside)
13. Autoclaved distilled water
14. *Taq* polymerase (1U/μl)
15. 10X *Taq* polymerase buffer with 1.5mM $MgCl_2$
16. 1X TAE buffer
17. Agarose
18. 6X Gel loading buffer
19. Solution B ($CHCl_3$: isoamylalcohol 24:1 v/v)

Method

1. Take 10μl (1μg) of M13mp18 (double stranded DNA) or pUC18 plasmid DNA in a sterile 1.5ml microcentrifuge tube and to that add 5μl 10X buffer.
2. Add 34μl of autoclaved distilled water.
3. Incubate the tube at 25 ^{0}C after adding 1μl of Sma I (10U/μl).

4. Repeat step 3 for two more times to ensure majority of the M13mp18 molecules are digested with Sma I.

5. Extract the M13mp18 DNA by adding equal volume of solution B and centrifuge at 10,000 rpm for 10 minutes at 4 ^{0}C.

6. Take out the clear aqueous phase in a sterile 1.5ml microcentrifuge tube and add equal volume of ice cold isopropanol and centrifuge at 10,000 rpm for 10 minutes at 4 ^{0}C.

7. Discard supernatant and add 500μl of 70% ethanol and perform centrifugation at 10,000 rpm for 10 minutes at 4 ^{0}C.

8. Discard the supernatant carefully. Air dry the pellet; dissolve the pellet in 50μl 10mMTris-HCl pH 8.

9. Take 5μl (0.5μg) of *Staphylococcus aureus* chromosomal DNA and in 50μl final volume add 100pmoles of forward and reverse primers, 5μl *Taq* 10X buffer and 1μl of *Taq* polymerase and perform polymerase chain reaction. The conditions are: initial denaturation step for 10 minutes at 94 ^{0}C; 40 cycles of 94 ^{0}C for 60 seconds, 46 ^{0}C for 60 seconds and for 100 seconds at 72 ^{0}C. Final extension step 72 ^{0}C for 5 minutes in a Mastercycler gradient Thermocycler (Eppendorf).

10. Resolve the PCR products by running 1% agarose gel electrophoresis in 1xTAE system at 100V for 45 minutes. In the gel, load molecular size markers in one well to evaluate the correct size of PCR product.

11. Cut the gel portion having correct size of PCR product and put it in a dialysis bag (which is prepared by boiling at 100 ^{0}C in a solution of 50mM of EDTA and 100mMNa$_2$CO$_3$) and add 2 ml of TE buffer (10mM Tris-HCl pH 8 and 10mM EDTA) and seal the dialysis bag with clips such that there is no leakage. Perform electrophoresis keeping the bag towards cathode for 90 minutes.

12. Take out the DNA solution from the dialysis bag and add equal volume of ice cold isopropanol and mix for 2 minutes in vortex mixer and perform centrifugation at 10,000 rpm for 10 minutes at 4 ^{0}C.

13. Discard the supernatant and add 500μl of 70% ethanol and carry out the centrifugation step.

14. Discard supernatant carefully and air dry the pellet and dissolve in 20μl of 10mMTris-HCl pH 8.

15. Take 10µl of PCR product of step 10, add 5µl of 10mM dNTP mix, 5µl of 10 X T4 DNA polymerase buffer, 29 µl of autoclaved distilled water and 1µl of T4 DNA polymerase and incubate at 37 ^{0}C for 1 hour.

16. Now add equal volume of solution B to the tube at step 11 and perform centrifugation at 10,000 rpm for 10 minutes at 4 ^{0}C.

17. Take out the aqueous phase cleanly in a sterile 1.5 ml microcentrifuge tube and add equal volume of ice cold isopropanol mix for 2 minutes and carry out centrifugation at 10,000 rpm for 10 minutes at 4 ^{0}C.

18. Discard the supernatant and add 250µl of 70% ethanol and perform centrifugation at 10,000 rpm for 10 minutes at 4 ^{0}C.

19. Discard the supernatant carefully and air dry the pellet; dissolve the pellet in 10µl of 10mM Tris-HCl pH 8.

20. Ligation reaction: In 1.5 ml sterile microcentrifuge tube add the following: 2µl of M13mp18 DNA of step 8, 5 µl of insert DNA of step 19, 2µl 10X blunt end ligation buffer, 10 µl autoclaved distilled water and 1 µl of (100U) T4 DNA ligase enzyme and keep it at 10 ^{0}C for overnight.

21. Transfect the ligation mixture to *E. coli* TG1 as described in 2.2.1 and screen the transparent plaques formed after 10h incubation at 37 ^{0}C.

22. With the help of a sterile tooth pick, take out one recombinant (clear plaque) from the plate by piercing into the plaque and put it in a 5ml LB tube in which *E. coli* TG1 bacteria are growing (mid log phase, i.e., OD_{540} = 0.5). Grow for another 6h at 37 ^{0}C.

23. Centrifuge the tube at 10,000 rpm for 10 minutes at 4 ^{0}C.

24. Take out the supernatant in a sterile 1.5ml tube and extract the single stranded DNA as described in chapter II. From the pellet extract the double stranded clone as described in chapter II.

25. Perform PCR as explained in step 9 using double stranded DNA of phage clone. This is to confirm the presence of insert in the phage DNA.

26. Take 5µl of single stranded DNA of step 24 in sterile 0.5ml PCR tube and to that add the following: 2.5µl of 10mM dNTP mix, 2.5µl of 10 X T7 DNA polymerase buffer, 5 µl (100pmoles) of mutated primer and 4 µl of autoclaved distilled water and 1µl of T7 DNA polymerase and incubate at 37 ^{0}C for 1 hour.

27. Perform transfection into *E. coli* TG1 bacteria as explained in chapter II.

28. Pick and isolate the clone as described in chapter II and step 22. Using the forward and reverse primers amplify the DNA as described in step 9.

Cloning and expression of mutated gene

29. The PCR product obtained in step is made into proper blunt ends following steps 15-19.

30. Take 5µl (1µg) pQE 30 DNA digest with Sma I enzyme following the steps 1-8.

31. Ligate the insert of step 29 and 30 following step 20. Here also take another tube which is called as ligation control in which insert is not added while Sma I digested vector is only present along with T4 DNA ligase, water and 10 x blunt end ligation buffer.

32. Perform transformation in *E. coli* DH5α as described in chapter 1.

33. Express the protein by inducing with IPTG as described in chapter 1.

34. Perform sonication to release the expressed protein in *E. coli* TG1 as described in chapter 1.

35. Purify the protein as described in chapter 1.

36. Analyze the protein by performing SDS-PAGE as explained in chapter 1 or if the gene encodes for an enzyme, perform its assay to know the effect of mutation and importance of that amino acid in the native protein.

References

1. Losick, Richard; Watson, James D.; Tania A. Baker; Bell, Stephen; Gann, Alexander; Levine, Michael W. (2008). *Molecular biology of the gene*. San Francisco: Pearson/Benjamin Cummings. ISBN 080539592X.

2. Russell, David W.; Sambrook, Joseph (2001). *Molecular cloning: a laboratory manual*. Cold Spring Harbor, N.Y:Cold Spring Harbor Laboratory. ISBN 0879695773.

3. Hartwell, Leland (2008). *Genetics: from genes to genomes*. Boston: Mc-GrawHill Higher Education. ISBN 0072848464.

4. Liang X, Peng L, Li K, Peterson T, Katzen F. A method for multi-site-directed mutagenesis based on homologous recombination. *Anal Biochem.* 2012;427(1):99-101.

5. Zhang BZ, Zhang X, An XP, Ran DL, Zhou YS, Lu J, Tong YG. An easy-to-use site-directed mutagenesis method with a designed restriction site for convenient and reliable mutant screening. *J Zhejiang Univ Sci B.* 2009;10(6):479-482.

6. Liu H, Naismith JH. An efficient one-step site-directed deletion, insertion, single and multiple-site plasmid mutagenesis protocol. *BMC Biotechnol.* 2008;8:91.

7. Edelheit O, Hanukoglu A, Hanukoglu I. Simple and efficient site-directed mutagenesis using two single-primer reactions in parallel to generate mutants for protein structure-function studies. *BMC Biotechnol.* 2009;9:61.

8. Yu Q. Cloning into Ml3 Bacteriophage Vectors. *Methods in Molecular Biology*, Vol 58 Basic DNA and RNA Protocols. Edited by A Harwood Humana Press Inc, Totowa, NJ.

9. Yarnsch-Perron C, Vieira J, Messing J. Improved Ml3 phage cloning vectors and host strains nucleotide sequences of the M13mp18 and pUC 19 vectors. 1985; *Gene* 33:103-119.

10. Messing J. New Ml3 vectors for cloning. *Meth. Enzymol.* 1983; 101: 20-79.

11. Walsh D. Cloning DNA Fragments in M13 Vectors. From: *The Nucleic Acid Protocols Handbook* Edited by: R. Rapley © Humana Press Inc., Totowa, NJ.

GENE REPLACEMENT EXPERIMENTS FOR THE CONSTRUCTION OF STABLE MUTANTS:

Background

A gene knockout is a genetic technique in which one of an organism's genes is made inoperative or removed from the organism. It is also termed as knockout organisms or simply knockouts. This technique is used to learn about a gene that has been sequenced, but whose function is either unknown or incompletely known. Researchers draw conclusions from the difference between the knockout organism and normal.

The availability of the genome sequences of several bacteria opens up new perspectives in the study of understanding the functions of genes. The artificial introduction of mutations, by molecular techniques, is a useful way of advancing our understanding of the genetics of bacteria. The method most commonly used to generate mutants involves integration of a plasmid into the chromosome by single crossover recombination. This method requires an internal fragment homologous to the target gene cloned into a vector carrying well defined genes such as resistance cassettes. This is a major limitation for systematic construction of mutants in post-genomic studies of any bacterial species. The possibility that a second crossover event will return the mutant to a wild-type phenotype is another important inconvenience. The gene replacement method is a useful way of overcoming these limitations (Figure 3).

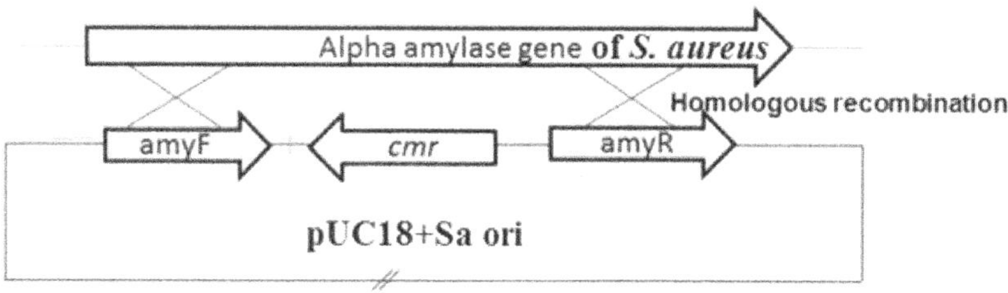

Figure 3   chematic representation of gene knockout strategy.

Materials

1. pUC18

2. Sma I (10U/µl) (NEB Biolabs)

3. *Staphylococcus aureus* ATCC12600

4. *Staphylococcus aureus* chromosomal DNA

5. Forward primer sarS1: ATGAATAAGCAATGGTGG and Reverse Primer sarS2: TTAATTTAATTCGATAACAC (100 pmoles/µl) of alpha-amylase gene of *Staphylococcus aureus*

6. Forward and Reverse primers (100 pmoles/µl) for amplification of *Staphylococcus aureus* origin (SaoriF: CAGGACATGCACCGC, SaoriR: AT-TATTGAGTCGAGT)

7. *E. coli* chromosomal DNA

8. 10mM dNTPs (dATP, dGTP, dTTP and dCTP)

9. Incubator 37 ^{0}C.

10. *E. coli* DH5α

11. pQE 30 vector

12. IPTG (Isopropyl βD-galactoside)

13. Autoclaved distilled water

14. *Taq* polymerase (1U/µl)

15. 10X *Taq* polymerase buffer with 1.5mM $MgCl_2$

16. T4 DNA ligase (100U/µl)

17. 10X Blunt end ligation buffer

18. 1 X TAE buffer

19. Agarose

20. 6X Gel loading buffer

21. Solution B (CHCl3: isoamylalcohol 24:1 v/v)

22. Forward primer CMF: ATGGAGAAAAAAATCACT and reverse primer CMR: TTACGCCCCGCCCTGCCA of Chloramphenicol acetylase gene of *E. coli.*

23. BsmB1 (10U/µl) (NEB Biolabs)

24. BsmB1 10 X restriction enzymes buffer

25. Calf intestine alkaline phosphatase (CIAP) (10U/µl)

26. TEG Buffer: 20% Glucose, 50mM Tris-HCl, pH to 8.0 and 50mM EDTA.

27. Lysis Buffer: 1% SDS, 0.2N NaOH

> Note: For preparing 10 ml of lysis buffer. Take 8.8 ml of sterile distilled water and to that add 0.2 ml of 10N NaOH and 1 ml of 10% SDS. (Note: Do not add SDS first as it will precipitate when NaOH is added to the solution).

28. 3M Potassium acetate pH 4.8: Dissolve 29.44 g potassium acetate in 25 ml of distilled water and adjust the pH to 4.8 with glacial acetic acid. Finally make up the volume to 100 ml of distilled water and sterilize by autoclaving at 15 lb / sq.in for 15 minutes.

29. RNase. Dissolve 10 mg / ml pancreatic RNase in 50mM Tris-HCl pH 7.5 containing 0.15M NaCl. Keep the above solution in boiling water bath for 10 minutes, this is to destroy the DNase function.

30. Isopropanol (ice cold): Chill the isopropanol by keeping it in ice before using.

31. 70% Ethanol: Mix v/v 70 ml of 100% ethanol and 30 ml of sterile distilled water.

32. 10 mg / ml Lysozyme solution. Dissolve 10 mg of lysozyme solution in 1 ml of TEG buffer and leave it on ice.

33. TE Buffer: 10mM Tris-HCl and 1mM EDTA, pH 8.0.

34. 100mg/ml Ampicillin solution: Dissolve 100mg of ampicillin in 1ml of autoclaved distilled water.

35. 50mg/ml Chloramphenicol: Dissolve 50mg of chloramphenicol in 1ml of pure ethanol.

36. T4 DNA polymerase (NEB Biolabs)

37. 10 X T4 DNA polymerase buffer

38. Bbv 1 (NEB Biolabs)

39. 10 X buffer for Bbv 1

40. 10% SDS

41. Solution C: Mix equal volume of solution A and solution B and store at room temperature [**Solution A**.: Phenol saturated with 1M Tris-HCl, pH 8 and stored in 0.1M Tris-HCl pH 8. **Solution B**: Mix Chororform and iso-amyl alcohol in 24: 1 (v/v)].

42. Brain Heart Infusion broth (Himedia)

43. LB agar plates: Dissolve 1g peptone, 0.5g yeast extract and 0.5g NaCl in about 80m of distilled water and adjust the pH to 7 with 10N NaOH, finally make up to 100ml with distilled water. Add 1.5g of agar powder to the above solution and autoclave at 121 ^{0}C for 20 minutes. To this

molten agar (at around 45 ^{0}C) appropriate antibiotic(s) are added and poured on to sterile plates in a laminar air flow.

Method

A. ***Construction of Shuttle plasmid***

    1. Polymerase chain reaction (PCR) to amplify *Staphylococcus aureus* origin of replication

1. Add 1µg (10µl) of *Staphylococcus aureus* chromosomal DNA in a PCR tube and add 1µl of forward (SaoriF) and reverse (SaoriR) primers for the amplification of 386bp DNA containing origin of replication. 2.5µl of 10X *Taq* polymerase buffer with 1.5mM MgCl$_2$, 1µl of dNTP mix, 8µl of sterile distilled water and 1µl of *Taq* DNA polymerase are also added.

2. Calibrate theromocycler (Eppendorf) with 94 ^{0}C for 10 minutes and 40 cycles of 94 ^{0}C for 1 minute (denaturation), 40 ^{0}C for 40 seconds (annealing), 72 ^{0}C for 45 seconds (amplification) and final extension at 72 ^{0}C for 5 minutes.

3. Perform 1.5% agarose gel electrophoresis to resolve the amplified products, along with PCR amplicons load molecular size markers in one well to evaluate the correct size of the PCR products. Excise the gel where appropriate molecular size PCR products are indicated and from the agarose gel electroelute the DNA molecules.

4. The agarose containing the PCR amplicons were added in a dialysis bag whose pore size is smaller than the size of the PCR products. One side of the dialysis bag is clipped with the help of dialysis tubing clips and then add TE buffer (10mM Tris-HCl pH 8, and 10mM EDTA) twice the volume of the agarose and seal the other end with the help of a dialysis tubing clip. Electrophoresis is carried out by keeping the dialysis bag towards cathode and after 45-60 minutes all the DNA molecules present in the agarose will be eluted into the buffer.

5. The DNA solution in the dialysis bag is carefully pippetted into a sterile 1.5ml microcentrifuge tube and add equal volume of ice cold isopropanol, mix the solution by vortexing; following this the tube is centrifuged at 10,000 rpm for 10 minutes at 4 ^{0}C.

6. Decant the supernatant carefully and dispense 250µl of 70% ethanol, after brief vortexing repeat centrifugation step as mentioned in step 5.

7. Discard the supernatant and air dry the tubes and dissolve the pellet in 20µl of 10mM Tris-Cl pH 8 buffer.

8. Take 10µl of PCR product of step 7, add 5µl of 10mM dNTP mix, 5µl of 10 X T4 DNA polymerase buffer, 29 µl of autoclaved distilled water and 1µl of T4 DNA polymerase and incubate at 37 ^{0}C for 1 hour.

9. To the above solution add equal volume of ice cold isopropanol and centrifuge at 10, 000rpm for 10 minutes at 4 ^{0}C.

10. Discard the supernatant and add 250µl of 70% ethanol and repeat the centrifugation step. Decant the supernatant and air dry the pellet. Suspend the pellet in 10µl of 10mM Tris-HCl pH 8, store the tube in -20 ^{0}C until further used.

2. BsmBI digestion of pUC 18 plasmid DNA

1. Take 10µl (1µg) of pUC18 plasmid DNA in a sterile 1.5ml microcentrifuge tube and add 5µl 10X buffer; further dispense 34µl of autoclaved distilled water, incubate the tube at 37 ^{0}C for 1h after adding 1µl of BsmBI (10U/µl). Repeat the incubation and enzyme addition two more times. This is to ensure majority of the molecules are cleaved by the restriction endonuclease. Following this to the same tube add 1 µl of CIAP and incubate at 37 ^{0}C for 1h, repeat the step one more time to ensure majority of the molecules are dephosphorylated by CIAP.

2. Extract the digested pUC18 DNA by adding equal volume of solution B and centrifugation at 10,000 rpm for 10 minutes at 4 ^{0}C.

3. Take out the clear aqueous phase in a sterile 1.5ml microcentrifuge tube and add equal volume of ice cold isopropanol and centrifuge at 10,000 rpm for 10 minutes at 4 ^{0}C.

4. Discard supernatant and add 250µl of 70% ethanol and perform centrifugation at 10,000 rpm for 10 minutes at 4 ^{0}C.

5. Discard the supernatant carefully, air dry the pellet; dissolve the pellet in 50µl 10mMTris-HCl pH 8. Thus obtained plasmid is named as pSa1.

3. Ligation of PCR amplicons and BsmBI digested pUC 18.

1. Ligation reaction: In 1.5 ml sterile microcentrifuge tube add the following: 2µl of pUC18 DNA of step 5, 5 µl of insert DNA of step 10, 2µl 10X blunt end ligation buffer, 10 µl autoclaved distilled water and 1 µl of (100U) T4 DNA ligase enzyme and keep it at 10 ^{0}C for overnight.

2. Similarly, prepare another tube in which all the ingredients of step 1 are added except insert DNA of step 10, instead of this add 5 µl of autoclaved distilled water and incubate at 10 ^{0}C for overnight.

4. Transformation into Staphyloccus aureus ATCC12600

1. Perform transformation in *Staphyloccus aureus* ATCC12600 as described in chapter 1 along with all controls with following modification: incubation of *Staphyloccus aureus* ATCC12600 in 0.1M $CaCl_2.2H_2O$ for 1h instead of 30 minutes, this is to ensure proper lysis of cell wall due to imbibing of hypotonic solution of $CaCl_2.2H_2O$.

2. Colonies appearing in the test plate contains shuttle pUC18 in which the *S. aureus* origin is inserted. To ensure the presence of shuttle plasmid in *S. aureus* ATCC12600 isolate the shuttle plasmid pSa1 as described below:

5. Isolation of pSa1 plasmid from Staphyloccus aureus ATCC12600

1. Grow Staphyloccus aureus ATCC12600 containing shuttle plasmid in 50µg/ml ampicillin at 37° C with a constant shaking at 150-200 rpm up to late log phase (OD540= 0.8 or 0.9)

2. Take 1.5 ml of the above grown culture in a sterile microcentrifuge tube and centrifuge the cells at 7,000 rpm for 90 seconds at 4 ^{0}C in a microcentrifuge.

3. Discard the supernatant and add 0.5 ml of TEG buffer and suspend the culture and centrifuge the cells at 7,000 rpm for 90 seconds at 4 ^{0}C in a microcentrifuge and perform this step thrice. This is to remove and inactivate all the DNase which is secreted in the culture filtrate of growing Staphyloccus aureus ATCC12600.

4. Discard the supernatant and suspend the pellet in 150µl of ice cold lysozyme solution and leave it at 37 ^{0}C for 120 minutes.

5. To the same above solution add 50 µl of 10mg/ml RNase and leave at 37 ^{0}C for 30 minutes.

6. Dispense 300 µl of lysis buffer and mix it carefully such that clear transparent solution is formed and leave it on ice for 10 minutes.

7. Now add 250 µl of 3M potassium acetate buffer pH 4.8 slowly and leave it on ice for 30 minutes.

8. Centrifuge at 10,000 rpm for 10 minutes at 4 ^{0}C, take out the supernatant in a sterile 1.5ml microcentrifuge tube and discard the pellet. To the supernatant add equal volume of ice cold isopropanol mix by gentle vortexing and leave it on ice for 2-3 minutes and repeat centrifugation step.

9. Discard the supernatant and add 0.5 ml of 70% ethanol and centrifuge at 10,000 rpm for 10 minutes at 4 ^{0}C, discard the supernatant and air dry the pellet.

10. Dissolve the pellet in 50µl of TE buffer. Analyze the plasmid DNA by running in one percent agarose gel, for this load molecular size marker in one well and in another intact pUC 18 plasmid DNA.

B. ***Cloning of alpha-amylase gene in pSa1 plasmid***

1. Polymerase chain reaction (PCR) to amplify Staphylococcus aureus alpha-amylase gene

1. Add 1µg (10µl) of Staphylococcus aureus chromosomal DNA in a PCR tube and add 1µl of forward (sarS1) and reverse (sarS2) primers for the amplification of alpha amylase gene. 2.5µl of 10X Taq polymerase buffer with 1.5mM MgCl2, 1µl of dNTP mix, 8µl of sterile distilled water and 1µl of Taq DNA polymerase are also added.

2. Calibrate theromocycler (Eppendrof) with 94 ^{0}C for 10 minutes and 40 cycles of 94 ^{0}C for 1 minute (denaturation), 33 ^{0}C for 60 seconds (annealing), 72 ^{0}C for 100 seconds (amplification) and final extension at 72 ^{0}C for 5 minutes.

3. Following PCR perform steps A. 1. 3-10 to get alpha-amylase encoding gene in pure form. Store the insert in -20 ^{0}C until further used.

2. Sma I digestion of pSa1 plasmid isolated in A. 5.

1. Take 10µl (1µg) of pSa1 plasmid DNA in a sterile 1.5ml microcentrifuge tube and add 5µl 10X Sma I buffer. Further dispense 34µl of autoclaved distilled water, incubate the tube at 25 ^{0}C for 1h after adding 1µl of Sma I (10U/µl). Repeat the incubation and enzyme addition two more times. This is to ensure majority of the molecules are cleaved by the restriction endonuclease. Following this, to the same tube add 1 µl of CIAP and incubate at 37 ^{0}C for 1h, repeat the step one more time to ensure majority of the molecules are dephosphorylated by CIAP.

2. Extract the digested pSa1 plasmid DNA by adding equal volume of solution B and perform centrifugation at 10,000 rpm for 10 minutes at 4 ^{0}C.

3. Take out the clear aqueous phase in a sterile 1.5ml microcentrifuge tube and add equal volume of ice cold isopropanol and centrifuge at 10,000 rpm for 10 minutes at 4 ^{0}C.

4. Discard supernatant and add 250µl of 70% ethanol and perform centrifugation at 10,000 rpm for 10 minutes at 4 ^{0}C.

5. Discard the supernatant, carefully air dry the pellet; dissolve the pellet in 50µl 10mMTris-HCl pH 8.

3. Ligation of Sma I digested pSa1 and alpha- amylase gene of S. aureus

1. Ligation reaction: In 1.5 ml sterile microcentrifuge tube add the following: 2µl of pSa1 DNA of step B 5, 5 µl of insert DNA of step B 3 , 2µl 10X blunt end ligation buffer, 10 µl autoclaved distilled water and 1 µl of (100U) T4 DNA ligase enzyme and keep it at 10 ^{0}C for overnight.

2. Similarly, prepare another tube in which all the ingredients of step 1 are added except insert DNA of step 10. Instead of this add 5 µl of autoclaved distilled water and incubate at 10 ^{0}C for overnight.

4. Transformation of pSa1 into E. coli DH5α

1. Perform transformation in E. coli DH5α as described in chapter 1 along with all controls.

2. Colonies appearing in the test plate contains pSa1 having S. aureus alpha-amylase gene. The plasmid is named as pSamy To ensure the presence of insert in pSa1 shuttle plasmid isolate the shuttle plasmid pSa1 as described below:

5. Isolation of pSamy plasmid from E. coli DH5α

1. Grow E. coli DH5α containing pSamy in 50μg/ml ampicillin at 37° C with a constant shaking at 150-200 rpm up to late log phase (OD540= 0.8 or 0.9).

2. Take 1.5 ml of the above grown culture in a sterile microcentrifuge tube and centrifuge the cells at 7,000 rpm for 90 seconds at 4 ^{0}C in a microcentrifuge.

3. Discard the supernatant and add 0.5 ml of TEG buffer and suspend the culture and centrifuge the cells at 7,000 rpm for 90 seconds at 4 ^{0}C in a microcentrifuge.

4. Discard the supernatant and suspend the pellet in 150μl of ice cold lysozyme solution and leave it at room temperature for 10 minutes.

5. To the same above solution add 50 μl of 10mg/ml RNase and leave at 37 ^{0}C for 30 minutes.

6. Dispense 300 μl of lysis buffer and mix it carefully such that clear transparent solution is formed and leave it on ice for 10 minutes.

7. Now add 250 μl of 3M potassium acetate buffer pH 4.8 slowly and leave it on ice for 30 minutes.

8. Centrifuge at 10,000 rpm for 10 minutes at 4 ^{0}C, take out the supernatant in a sterile 1.5ml microcentrifuge tube and discard the pellet. To the supernatant add equal volume of ice cold isopropanol mix by gentle vortexing and leave it on ice for 2-3 minutes and repeat centrifugation step.

9. Discard the supernatant and add 0.5 ml of 70% ethanol and centrifuge at 10,000 rpm for 10 minutes at 4 ^{0}C. Discard the supernatant and air dry the pellet.

10. Dissolve the pellet in 50µl of TE buffer. Analyze the pSamy plasmid DNA by running in one percent agarose gel, for this load molecular size marker in one well and in another intact pSa1 plasmid DNA.

6. Digestion of pSamy plasmid with Bbv1

1. Take 10µl (1µg) of pSamy plasmid DNA in a sterile 1.5ml microcentrifuge tube and add 5µl 10X Bbv I buffer. Further dispense 34µl of autoclaved distilled water, incubate the tube at 37 ^{0}C for 1h after adding 1µl of Bbv I (10U/µl). Repeat the incubation and enzyme addition two more times, this is to ensure majority of the molecules are cleaved by the restriction endonuclease. Following this, to the same tube add 1 µl of CIAP and incubate at 37 ^{0}C for 1h, repeat the step one more time to ensure majority of the molecules are dephosphorylated by CIAP.

2. Extract the digested pSamy plasmid DNA by adding equal volume of solution B and perform centrifugation at 10,000 rpm for 10 minutes at 4 ^{0}C.

3. Take out the clear aqueous phase in a sterile 1.5ml microcentrifuge tube and add equal volume of ice cold isopropanol and centrifuge at 10,000 rpm for 10 minutes at 4 ^{0}C.

4. Discard supernatant and add 250µl of 70% ethanol and perform centrifugation at 10,000 rpm for 10 minutes at 4 ^{0}C.

5. Discard the supernatant carefully, air dry the pellet; dissolve the pellet in 50µl 10mMTris-HCl pH 8.

7. Polymerase chain reaction (PCR) to amplify E. coli chloramphenicol acetyltransferase (cat) gene

1. Add 1µg (10µl) of E. coli chromosomal DNA in a PCR tube and add 1µl of forward (CMF) and reverse (CMR) primers for the amplification of cat. 2.5µl of 10X Taq polymerase buffer with 1.5mM MgCl2, 1µl of dNTP mix, 8µl of sterile distilled water and 1µl of Taq DNA polymerase are also added.

2. Calibrate theromocycler (Eppendrof) with 94 ^{0}C for 10 minutes and 40 cycles of 94 ^{0}C for 1 minute (denaturation), 49 ^{0}C for 60 seconds (annealing), 72 ^{0}C for 100 seconds (amplification) and final extension at 72 ^{0}C for 5 minutes.

3. Following PCR perform steps A. 1. 3-10 to get cat encoding gene in pure form. Store the insert in -20 ^{0}C until further use.

8. Ligation of Bbv I digested pSamy and cat gene of E. coli

1. Ligation reaction: In 1.5 ml sterile microcentrifuge tube add the following: 2µl of pSamy DNA of step B 6, 5 µl of insert DNA of step B 7 , 2µl 10X blunt end ligation buffer, 10 µl autoclaved distilled water and 1 µl of (100U) T4 DNA ligase enzyme and keep it at 10 ^{0}C for overnight.

2. Similarly, prepare another tube in which all the ingredients of step 1 are added except insert DNA of step 10, instead of this add 5 µl of autoclaved distilled water and incubate at 10 ^{0}C for overnight.

9. Transformation into Staphyloccus aureus ATCC12600

1. Perform transformation in Staphyloccus aureus ATCC12600 as described in A 4 along with all controls with following modification: incubation of Staphyloccus aureus ATCC12600 in 0.1M $CaCl_2.2H_2O$ for 1h instead of 30 minutes, this is to ensure proper lysis of cell wall due to imbibing of hypotonic solution of $CaCl_2.2H_2O$.

2. Colonies appearing in the test plate contains pSamy in which the cat gene is inserted. To ensure the presence of insert in pSamy, in all LB agar plates include chloramphenicol (50µg/ml) along with 100µg/ml ampicillin. Isolate the plasmid pSamycat containing cat gene.

10. Isolation of pSamycat plasmid from Staphyloccus aureus ATCC12600

1. Grow *Staphyloccus aureus* ATCC12600 containing pSamycat in 50µg/ml ampicillin and 25 µg/ml chloramphenicol at 37° C with a constant shaking at 150-200 rpm up to late log phase (OD540= 0.8 or 0.9).

2. Take 1.5 ml of the above grown culture in a sterile microcentrifuge tube and centrifuge the cells at 7,000 rpm for 90 seconds at 4 ^{0}C in a microcentrifuge.

3. Discard the supernatant and add 0.5 ml of TEG buffer and suspend the culture and centrifuge the cells at 7,000 rpm for 90 seconds at 4 ^{0}C in a microcentrifuge and perform this step thrice. This is to re-

move and inactivate all the DNase which is secreted in the culture filtrate of growing *Staphyloccus aureus* ATCC12600.

4. Discard the supernatant and suspend the pellet in 150µl of ice cold lysozyme solution and leave it at 37 ^{0}C for 120 minutes.

5. To the same above solution add 50 µl of 10mg/ml RNase and leave at 37 ^{0}C for 30 minutes.

6. Dispense 300 µl of lysis buffer and mix it carefully such that clear transparent solution is formed and leave it on ice for 10 minutes.

7. Now add 250 µl of 3M potassium acetate buffer pH 4.8 slowly and leave it on ice for 30 minutes.

8. Centrifuge at 10,000 rpm for 10 minutes at 4 ^{0}C, take out the supernatant in a sterile 1.5ml microcentrifuge tube and discard the pellet. To the supernatant add equal volume of ice cold isopropanol mix by gentle vortexing and leave it on ice for 2-3 minutes and repeat centrifugation step.

9. Discard the supernatant and add 0.5 ml of 70% ethanol and centrifuge at 10,000 rpm for 10 minutes at 4 ^{0}C. Discard the supernatant and air dry the pellet.

10. Dissolve the pellet in 50µl of TE buffer. Analyze the plasmid DNA by running in one percent agarose gel, for this load molecular size marker in one well and in another intact pSamy plasmid DNA.

C. ### *Screening of cat gene incorporation in S. aureus ATCC12600*

1. Correct homologous recombination of the target region was established by PCR using primers sarS1 and sarS2 in pSamy (plasmid) and chromosomal DNA of *S. aureus* ATCC12600 containing pSamycat plasmid. Follow C2 procedure and the PCR conditions mentioned in B7.

C 2. For isolation of chromosomal DNA from S. aureus containing pSamycat plasmid.

i. Grow *Staphyloccus aureus* ATCC12600 containing pSamycat in 50µg/ml ampicillin and 25 µg/ml chloramphenicol at 37° C with a constant shaking at 150-200 rpm up to late log phase (OD540 = 0.8 or 0.9).

ii. Take 1.5 ml of the above grown culture in a sterile microcentrifuge tube and centrifuge the cells at 7,000 rpm for 90 seconds at 4 ^{0}C in a microcentrifuge.

iii. Discard the supernatant and add 0.5 ml of TEG buffer and suspend the culture and centrifuge the cells at 7,000 rpm for 90 seconds at 4 ^{0}C in a microcentrifuge and perform this step thrice. This is to remove and inactivate all the DNase which is secreted in the culture filtrate of growing *Staphyloccus aureus* ATCC12600.

iv. Discard the supernatant and suspend the pellet in 150μl of ice cold lysozyme solution and leave it at 37 ^{0}C for 120 minutes.

v. To the same above solution add 50 μl of 10mg/ml RNase and leave at 37 ^{0}C for 30 minutes.

vi. Dispense 100 μl of 10% SDS solution and keep at 37 ^{0}C for 120 minutes.

vii. Now add equal volume of Solution C and vortex vigorously for 2 minutes and centrifuge at 10,000 rpm for 10 minutes at 4 ^{0}C, take out the supernatant in a sterile 1.5ml microcentrifuge tube and add equal volume of ice cold isopropanol. Mix by gentle vortexing and leave it on ice for 2-3 minutes and repeat centrifugation step.

viii. Discard the supernatant and add 0.5 ml of 70% ethanol and centrifuge at 10,000 rpm for 10 minutes at 4 ^{0}C. Discard the supernatant and air dry the pellet.

ix. Dissolve the pellet in 50μl of TE buffer. Analyze the chromosomal DNA by running in one percent agarose gel.

2. Grow the integrants at 37 ^{0}C overnight with shaking in 10 ml BHI broth (Himedia) without any antibiotics. Inoculate 100 μl of the grown culture in 10 ml fresh BHI broth and grow up to late log phase at 37 ^{0}C. Repeat this step for two more times, the last step is growing S. aureus bacteria are plated in LB agar plates containing 100μg/ml ampicillin. If the plasmid is cured from the recombinants then the bacteria will not be able to grow in the presence of ampicillin as S. aureus ATCC12600 is sensitive to ampicillin.

3. If no growth of the integrants in ampicillin is observed, then plate the bacteria in LB agar plates containing 50µg/ml chloramphenicol and incubate overnight at 37 ^{0}C. The colonies observed in the chloramphenicol plate indicate integration of cat gene in the chromosome of S. aureus ATCC12600.

References

1. Aranda J, Poza M, Pardo BG, Rumbo S, Rumbo C, Parreira JR, Rodríguez-Velo P, Bou G. A rapid and simple method for constructing stable mutants of *Acinetobacter baumannii*. *BMC Microbiol*. 2010;10:279.

2. Kato F, Sugai M. A simple method of markerless gene deletion in *Staphylococcus aureus*. *J Microbiol Methods*. 2011;87(1):76-81.

3. Kim SB, Timmusk S. A Simplified Method for Gene Knockout and Direct Screening of Recombinant Clones for Application in *Paenibacillus polymyxa*. *PLoS One*. 2013;8(6):e68092.

4. Geng SZ, Jiao XA, Pan ZM, Chen XJ, Zhang XM, Chen X. An improved method to knock out the asd gene of *Salmonella enterica* serovar Pullorum. *J Biomed Biotechnol*. 2009;2009:646380.

5. Nakashima N, Miyazaki K. Bacterial cellular engineering by genome editing and gene silencing. *Int J Mol Sci*. 2014;15(2):2773-2793.

6. Xu Z, Zikos D, Osterrieder N, Tischer BK. Generation of a complete single-gene knockout bacterial artificial chromosome library of cowpox virus and identification of its essential genes. *J Virol*. 2014;88(1):490-502.

7. Baba T, Bae T, Schneewind O, Takeuchi F, Hiramatsu K. Genome sequence of Staphylococcus aureus strain Newman and comparative analysis of staphylococcal genomes: polymorphism and evolution of two major pathogenicity islands. *J Bacteriol*. 2008;190(1):300-310.

8. Stark, MW, Sherratt, D, Boocock, MR (1989). Site-specific recombination by Tn 3 resolvase:topological changes in the forward and reverse reactions. *Cell*. 58 (4): 779–90.

9. Russell, David W.; Sambrook, Joseph (2001). *Molecular cloning: a laboratory manual*. Cold Spring Harbor, N.Y:Cold Spring Harbor Laboratory. ISBN 0879695773.

10. Potrykus J, Wegrzyn G. Chloramphenicol-sensitive *Escherichia coli* strain expressing the chloramphenicol acetyltransferase (cat) gene. *Antimicrob Agents Chemother*. 2001;45(12):3610-3612.

11. Lakshmi HP, Prasad UV, Yeswanth S, Swarupa V, Prasad OH, Narasu ML, Sarma PV. Molecular characterization of α-amylase from *Staphylococcus aureus*. *Bioinformation*. 2013; 9(6): 281–285.

12. Leonard AC, Mechali M. *DNA Replication Origins*. Cold Spring Harb Perspect Biol 2013;5:a010116.

13. Grkovic S, Brown MH, Hardie KM, Firth N, Skurray RA. Stable low-copy-number *Staphylococcus aureus* shuttle vectors. *Microbiology*. 2003;149(Pt 3):785-794.

14. Rajewska M, Wegrzyn K, Konieczny I. AT-rich region and repeated sequences - the essential elements of replication origins of bacterial replicons. *FEMS Microbiol Rev*. 2012;36(2):408-434.

15. Badrinarayanan A, Le TB, Laub MT. Bacterial chromosome organization and segregation. *Annu Rev Cell Dev Biol*. 2015;31:171-199.

GLOSSARY

Allele: Alternative forms of a gene, for example, in a normal gene sequence ......ATT CTG GCC CAT.... in the codon CTG, G is replaced with T, that gene is the allele of the normal gene.

Antibodies: *In vivo* generation of immunoglobulin molecules against epitopes of an antigen are called antibodies.

Bacterial Artificial Chromosome (BAC): A bacterial artificial chromosome (BAC) is generated by using the F-plasmid which exists as single copy in the host *E. coli* due to the presence of partition genes (*sopA*, *sopB* and *sopC*) and cell division control gene which controls the division of host *E. coli* and the insert DNA size is usually around 150-350 kbp.

Bacteria: One of the smallest living organisms of the planet, has a thick permeable cell wall made up of peptidoglycan and cell membrane wherein the organism carries out all the metabolic functions utilizing the precursors present in the environment and divides steadily to establish itself in that niche.

cDNA Library: The group of bacterial colonies formed upon transformation of recombinant DNA molecules generated by the ligation of complementary DNA (cDNA) molecules which are the mRNAs of a cell converted into cDNA by reverse transcriptase enzyme.

Chromosome: DNA wound on a histone octamer protein present in the nucleus of a cell which can be observed by staining with nuclear specific stains like Giemsa Stain or Leishman stain, etc., as deep blue bands when observed under microscope is called as chromosome.

Codon: A three nucleotide sequence present on mRNA representing an amino acid is called as codon. A codon also has a definite opposite sequence present on the tRNA in the anitcodon arm.

Colony Hybridization: The cloned gene in a plasmid vector and transformed into competent bacteria and further identification of colony/colonies among all the transformants using the gene specific probe is called colony hybridization.

Conjugation: It is the process of transfer of plasmid DNA from donor bacteria to recipient bacteria through conjugation tube in case of Gram negative bacteria while in Gram positive bacteria it is through surface to surface contact. The movement of plasmid DNA is promoted by series of genes traA, traL, traE, traJ, traI, traO, traM traJ and traL present in donor and recipient bacteria.

Cos Site: Lambda phage DNA has 12 base pair cohesive ends which are called as cos sites. On entry into the host *E. coli*, the phage DNA immediately circularises and forms covalently closed circular DNA using the cos sites (**Cos site Left GGGCGGCGACCT and Cos site Right CCCGC-CGCTGGA**), and this reaction is catalyzed by host DNA ligase.

DNA Ligases: This enzyme present in bacteria catalyzes the formation of phosphodiester bond between nucleotides in a double stranded DNA using NADH as cofactor.

DNA Polymerase: An enzyme which catalyzes formation of phosphodiester bonds between nucleotides taking cues from a template DNA strand to which an oligonucleotide primer is attached which has free 3'OH. The polymerase action is always in the 5' to 3'direction while forming daughter strand will be in the 3' to 5' direction.

DNA: Deoxyribonucleic acid is made up of two polynucleotide chains running in opposite direction with respect to deoxyrbiose linked together by hydrogen bonds in definite pattern; Adenine forms two hydrogen bonds with thymine and Guanine forms three hydrogen bonds with cytosine. Nucleotides in DNA are linked together by phosphodiester bonds between 5'phosphate of one nucleotide and 3'OH of another nucleotide. As both the strands are helical in nature, upon hydrogen bonding it gives rise to a double helical structure and this bonding pattern is conserved in entire biotic community.

Electrophoresis: Separation of proteins or nucleic acids by applying electricity using a matrix is called electrophoresis.

Endonucleases: Enzymes that hydrolyses the phosphodiester bond within the double stranded DNA are called as endonucleases.

Expression Vector: A replicon carrying promoter, ribosome binding site, unique type II restriction endoculease sites, transcription terminator and marker gene is called as expression vector.

F-plasmid: Fertility factor or plasmid present in *E. coli* makes the organism to participate in the conjugation phenomenon. This plasmid has the capacity to integrate with the chromosome of *E. coli* forming high frequency (Hfr) recombinants. This F plasmid has the following crucial genes: *sopA*, *sopB* and *sopC* genes which help the plasmid to exist as single copy in the host *E. coli*. It also has *tra* and *mob* genes—these genes participate in the conjugation phenomenon. Integration region where the plasmid participates in the integration with the chromosome of *E. coli* and *cell division control* gene which controls the division of *E. coli*.

Gam genes: Gam genes are present on the lambda phage DNA whose products helps in the conversion of phage DNA to replicate in rolling circle mode whereby the formed DNA can be directly packaged into phage heads.

Gene: A DNA sequence having two regulatory sequences, the promoter and terminator, and coding region (s) is referred as gene, for example: alpha-amylase gene encodes single coding controlled from single promoter and terminator.

Genome: Total DNA content of an organism is called genome.

Genomic Library: A pool of bacterial colonies carrying the genes of an organism is called genomic library.

Gram Staining: The bacteria are usually stained with primary stain called Crystal Violet and fixed with a mordant iodine, some of the bacteria are able to retain the primary stain even after treatment with alcohol while some are decolorised on adding alcohol. The bacteria which retain crystal violet even after treatment with decolorising agent are referred as Gram positive bacteria while the bacteria that do not retain the stain are called as Gram negative bacteria. This is primarily due to the thick cell wall made up of peptidoglycan present in Gram positive bacteria that retains the crystal violet while in Gram negative bacteria the peptidoglycan layer is very thin and therefore gets easily replaced by alcohol, decolorised upon treatment with alcohol. These are then counter stained with safranin and are observed as pink coloured bodies under microscope. This stain and method was developed by **Danish bacteriologist Hans Christian Gram** hence the name Gram straining after this great scientist.

Heterogenous RNA (hnRNA):   The RNA generated from a gene in the nucleus by RNA polymerase II is called hnRNA. This molecule contains non-coding sequences called introns and coding sequences called exons.

Immunoblot: Identification of proteins that are transferred electrophoretically on the nitrocellulose membrane using either monoclonal or polyclonal antibodies is called as immunoblot.

Immunological Screening: Screening of a genomic or cDNA library using polyclonal or monoclonal antibodies is called as immunological screening.

Lambda Phage: Lambda phage is a temperate phage that infects *E. coli* and is about 48.5Kbp in size. This phage has two phases of life cycle: one phase where the genome gets integrated in the genome of the host *E. coli* called the lysogenic state and second it propagates into number of copies in the host, results in the lysis of the host and releases as phages in the culture lysate. This phage genome is manipulated largely in the genes encoding for lysogenic state for the construction of vectors. These vectors are routinely used in the construction of genomic and cDNA libraries.

M13 Phage: M13 phage is a filamentous phage; M stands for Munich city where this viral particle was first discovered. This phage infects *E. coli* strains having F-plasmid; where it binds to the tra A gene product (conjugation tube) through which the phage enters the host *E. coli*. Here the phage slows down the growth of the host *E. coli* when the phage is propagating within the host, and the formed phages are then thrown into the culture filtrate of the host *E. coli*. The genome of this phage is single stranded DNA which is covalently closed circular in nature comprising of 6409bp and is encapsulated by capsid proteins of gene VIII and gene III of the phage.

Messenger RNA (mRNA): The product formed by the transcription of a gene by RNA polymerase is called messenger RNA. The same in eukaryotes is first formed as heterogenous nuclear RNA (containing both introns and exons) in the nucleus of a cell where it undergoes post transcriptional modification (removal of introns and joining of exons) to form mRNA which is transported to cytoplasm to get translated into a polypeptide molecule.

Monoclonal Antibody: Antibodies against single epitope of an antigen are called as monoclonal antibodies.

Nickel Metal Chelate Chromatography: This is affinity chromatography technique in which the matrix is tagged with Ni ions and through this matrix a protein solution in which proteins having continuous stretch of six histidine residues either in the NH_2-terminal or in the COOH will bind with Ni+ ions through coordinate covalent bond. Thus, bound molecules can be eluted using chaotropic reagents (Guanidium hydrochloride or Imidazole hydrochloride) which can be easily removed through dialysis. Recombinant proteins formed on cloning gene(s) in pQE 30 expression vectors or in pET vectors can be easily purified through this technique.

Nucleoside: Nitrogenous bases purines or pyrimidines linked with either deoxy ribose or with ribose through N-glycosidic bond.

Nucleotide: A nucleoside containing phosphate group at the 5th carbon position of deoxy ribose or in the ribose.

Open Reading Frame: A sequence of nucleotides continuously encoding for a minimum of 50 amino acids is called open reading frame.

Plaque Hybridization: The cloned gene in a phage vector and transfected into competent bacteria and further identification of plaque/plaques among all the transformants using the gene specific probe is called plaque hybridization.

Plasmids: Extrachromosomal replicons existing in the cytoplasm of a variety of bacteria possess unique character(s) which plays important role in facilitating the bacteria to grow effectively in the specified niche. Like bacterial chromosomal DNA, plasmid DNA is also covalently closed circular DNA.

Polyacrylamide Gel Electrophoresis (PAGE): Acrylamide monomer gets attached to another monomer with the help of bis-acrylamide and this process is carried out in the presence of ammonium per sulphate and Tetramethylethylenediamine (TEMED) which acts as free radical stabilizer. Thus formed polyacrylamide gel exhibiting molecular sieving (decreasing pore sizes from top to bottom) property is used as matrix in the electrophoresis for the separation of proteins or nucleic acids.

Polyclonal Antibody: Antibodies against all the epitopes of an antigen are called as polyclonal antibodies.

Polymerase Chain Reaction (PCR): This is a technique which utilizes the ability of *Taq* DNA polymerase to remain active even at 95 ^{0}C. Here a template DNA is taken and primers for both the strands in the 5' to 3' direction are added, and optimal buffering conditions, deoxy nucleotide mixture are provided and supplemented with Mg^{2+} ions that keeps the DNA in B-form of conformation which is the substrate for polymerase enzyme. The temperature is raised to 95 ^{0}C where the DNA is denatured and converted into single strands due to breaking of hydrogen bonds between bases, this is followed by lowering the temperature to that point for a brief time where only primers anneal with the template strands and this followed by the amplification at 72 ^{0}C for a brief time after which it goes back to the first step of denaturation at 95 ^{0}C and this cycle continues for 35 to 40 cycles. Then virtually from a single template DNA molecule a billion molecules can be amplified. As this is carried out in a cyclic manner, this is termed as polymerase chain reaction.

Polypeptide Chain: A polymer of amino acids joined together in the form of peptide bond is called polypeptide chain.

Probe: A DNA or RNA sequence attached with radioactive isotope or fluorescent dye or with biotin which can identify its complementary sequence in the pool of DNA fragments is called probe.

Proteome: Total proteins produced in the organism in its life time are called proteome.

Purines: Adenine and Guanine are purines

Pyrimidines: Cytosine, Thymine and Uracil are Pyrimidines

Replicon: A DNA molecule having origin sequence is called replicon, at this origin sequence the DNA polymerase initiates replication under *in vivo* conditions.

Restriction Endonucleases (RE):   Endonucleases that cleave phosphodiester bond only at a specific sequence in a double stranded DNA are called as restriction endonucleases. There are three types of restriction endonucleases classified on the mode of hydrolysis of phosphodiester bond in a double stranded DNA named as type I RE, type II RE and type III RE. Type I RE are big complex molecules, require S-adenosyl methonine, ATP and Mg^{2+} ions for the catalytic function and they hydrolyze DNA sequence minimum 1000bp away from the recognition sequence, similarly type III RE also requires same molecules for their catalytic function however they hydrolyze the DNA at about 25 to 30bp away from the recognition sequence. Type II RE requires only Mg^{2+} ions for their endonucleolytic function and they hydrolyze double stranded DNA within the recognition sequence only.

Ribosomes: Supramolecular organization of RNA and proteins is called as ribosomes.

RNA: Ribonucleic acid is also a single polynucleotide chain; here, the nucleotides are made up of ribose and instead of thymine here they have uracil which also forms two hydrogen bonds with adenine. Like DNA, this pattern is also conserved in biotic community.

Sodium Dodecyl Sulphate Polyacrylamide Gel Electrophoresis (SDS-PAGE):   Polyacrylamide gel matrix, loading buffer and running buffer containing SDS together are called as sodium dodecyl sulphate polyacrylamide gel electrophoresis (SDS-PAGE). This method of electrophoresis is chiefly carried out for the characterization of proteins, molecular weight determination, and monomeric nature of the protein. Here the intrinsic charge on the protein is masked by SDS and gives definite shape, thereby, the charge to mass ratio becomes constant for all the proteins that are separated in the gel while tertiary, secondary structures and covalent bonds are broken by boiling the protein sample mixed with loading buffer containing β-mercaptoethanol. Apparently, it appears that each protein now separates based on its molecular weight due to the molecular sieving property of the polyacrylamide gel.

T4 DNA Ligase: This enzyme is present in the T4 bacteriophage which infects *E. coli.* This enzyme also catalyzes the formation of phosphodiester bond between nucleotides in a double stranded DNA using ATP as cofactor.

T4 DNA Polymerase: An enzyme present on T4 bacteriophage that has polymerase activity in the 5' to 3' direction also has an exonuclease activity in the 3' to 5' direction. This exonuclease activity is always masked by the polymerase activity.

Taq **DNA Polymerase**: This is also a polymerase enzyme isolated from thermophilic bacteria called *Thermus aquaticus*. It has DNA polymerase whose optimal enzyme activity was observed at 72 ^{0}C and was able to survive even if it is kept at 94 ^{0}C. This property of this enzyme led to the development of polymerase chain reaction in 1983 by Kary B.Mullis and his co investigators. Like all other DNA polymerases this enzyme also has 5' to 3' polymerase activity and 3' to 5' exonuclease activity.

Thermus Aquaticus: A thermophilic bacteria which grows at temperature range of 50 ^{0}C to 80 ^{0}C. First reported by Thomas Brock who isolated this bacteria from Yellowstone national park of USA. Brock and Edwards (1970) reported on the fine structure *of Thermus aquaticus*. Degryse et al. (1978) and Degryse and Glansdorff (1981) studied the metabolism of *Thermus aquaticus*. These bacteria are obligately aerobic and chemoheterotrophic (Degryse et al. 1978).

Transcriptome: Total RNA synthesized in the organism in its life time is called transcriptome.

Transduction: Transfer of genetic material from one bacterium to another bacterium with the help of a bacteriophage is called transduction. There are two types of transduction: **Generalized transduction** in which any bacterial gene may be transferred to another bacterium via a bacteriophage, this is a very rare phenomenon and occurs in 1 phage in 10,000. **Specialized transduction** is the process by which a limited position of bacterial genes is transferred to another bacterium by the phage. The genes that get transferred (donor genes) depend on where the phage genome is located on the chromosome of the bacteria. Here, the prophage excises indistinctly from the chromosome such that bacterial genes lying nearby to the prophage are incorporated in the excised DNA which is then packaged into bacteriophage. This on infecting another bacterium introduces the incorporated DNA into the genome thus causing variations in the recipient bacterial genome.

Transfection: Transfection is referred as intentional introduction of DNA into eukaryotic cells. In this process, DNA or RNA are first encapsulated into liposomes and then when incubated with growing eukaryotic cells, these liposomes fuse with the cell membrane and thus deliver the DNA content into the cells. This process can also be carried out with viruses whose genome is recombined with desired DNA and these recombinant viral particles on incubation with eukaryotic cells results in the transfer of DNA into the cells.

Transformation: Direct uptake of exogenous DNA by the bacteria is called transformation. This is experimentally carried out in bacteria growing in mid-logarithmic phase with the help of calcium chloride or with the help of electric impulse. In both these methods, the bacterium is first made cell wall less and then the DNA is introduced through the cell membrane.

Translation: Conversion of mRNA sequence into amino acid sequence with the help of ribosomes and transfer RNA to generate a polypeptide chain is called translation.

Vector: A replicon which can carry foreign DNA is called vector, for this to happen the vector must have expressible trait (s) with unique restriction endonuclease site in it and this site can be used to break the DNA and ligate foreign DNA.

Western Blot: Electrophoretic transfer of resolved proteins either on PAGE or SDS-PAGE to a nitrocellulose membrane in an alkaline buffer containing 10% methanol is called as western blot.

Alpha Peptide: First 146 amino acids of β-galactosidase is called as alpha peptide.

β-galactosidase: This is an enzyme which hydrolyses lactose to glucose and galactose. This enzyme is the part of lactose operon present in *E. coli.*

Operon: More than one cistron controlled from the same promoter is called as operon.

INDEX

O

Oligo dT   21, 67, 73

P

PCR products   15, 106, 109, 127, 134

Plasmid DNA   2, 3, 4, 8, 10, 19, 38, 56, 71, 103, 126, 135, 137, 138, 140, 142, 148, 152

Polyclonal Antibodies   27, 28, 29, 81, 150, 152

Polyethylene Glycol   55

Polymerase Chain Reaction (PCR)   15

Potassium Acetate   8, 9, 10, 41, 54, 56, 94, 95, 133, 137, 139, 142

Protein   27, 28, 29, 31, 34, 35, 36, 37, 38, 42, 58, 59, 60, 81, 83, 85, 86, 117, 118, 126, 129, 130, 147, 151, 153, 154

R

Recombinant DNA   57, 87, 147

S

Sodium Dodecyl Sulfate   8

Streptomyces   51

Sucrose   31, 40, 90, 91, 92, 98, 101

T

T7 RNA Polymerase   3

Taq DNA Polymerase   104, 120, 134, 137, 140, 152

TE Buffer   10, 18, 20, 43, 44, 45, 46, 47, 56, 57, 65, 70, 71, 91, 95, 96, 98, 106, 111, 112, 127, 134, 137, 140, 142, 143

TEG Buffer   8, 54, 94, 132

Tetrahydrochloride   26, 30, 36, 80, 82, 122

Toxin Assay   37

V

Virus   51, 58, 91, 144

W

Water Molecules   10

Y

Yeast Extract   53, 63

Made in the USA
Monee, IL
07 July 2026